The Genetic Revolution and Human Rights

The Genetic Revolution and Human Rights

The Oxford Amnesty Lectures 1998

EDITED BY

Justine Burley

OXFORD
UNIVERSITY PRESS

OXFORD
UNIVERSITY PRESS

Great Clarendon Street, Oxford OX2 6DP

Oxford University Press is a department of the University of Oxford.
It furthers the University's objective of excellence in research, scholarship,
and education by publishing worldwide in

Oxford New York

Athens Auckland Bangkok Bogota Bombay Buenos Aires Calcutta
Cape Town Chennai Dar es Salaam Delhi Florence Hong Kong Istanbul
Karachi Kuala Lumpur Madrid Melbourne Mexico City Mumbai
Nairobi Paris Sao Paulo Singapore Taipei Tokyo Toronto Warsaw

with associated companies in Berlin Ibadan

Oxford is a registered trade mark of Oxford University Press
in the UK and in certain other countries

Published in the United States
by Oxford University Press Inc., New York

British Library Cataloguing in Publication Data
Data available

Library of Congress Cataloging in Publication Data
Data available

ISBN 0-19-286201-4

5 7 9 10 8 6 4

Phototypeset in Bembo
by Intype London Ltd
Printed in Great Britain by
Clays Ltd, St Ives plc

Foreword

Richard Dawkins

The Oxford Amnesty lectures have established themselves as a respected institution, even tradition, with commendable speed. The tradition began as recently as 1992, but already its prestige makes an invitation to deliver an Amnesty lecture harder to resist than most other invitations a scholar might receive. Perhaps this owes a little to the aesthetic appeal of Christopher Wren's Sheldonian Theatre, obviously much more to Amnesty International's unique moral standing in the world, and then something to the distinction of the company one would be keeping.

Tradition it may have become but, like other good traditions, the Amnesty Lecture Series is changeable. In 1997 the organizers chose for the first time a scientific theme, The Values of Science. Dr Wes Williams, the Editor of the resulting volume (and one of the founders of the Amnesty Lectures themselves), has explained that these lectures

generated such interest that the organisers broke the rules. They decided to pursue the same theme—science and its relation to rights—two years running. The flurry of questions, moral panic and real concern that was generated by the publicity surrounding the cloning of Dolly the sheep made it clear not only that important advances were, and are, being made in scientific inquiry and technology, but also that we need both informed scientists and clear-headed people to join in reasoned debate as to the implications of this 'revolution' for society.

In that previous Amnesty series, Nicholas Humphrey began his excellent lecture, 'What shall we tell the children?', by quoting—and dissenting from—the proverb 'Sticks and stones may break my bones, but words will never hurt me.' The Watson/Crick revolution has shown us that DNA is all words. Genes are digitally coded text, in a sense too full to be dismissed as analogy. Like human words they have the power to hurt, and that power is the greater because, given the right conditions, DNA words can dictate with stronger predictability than most human imperatives. The possessor of the late-acting dominant gene Huntington's Chorea can anticipate a future so bleak that many would prefer to be kept in ignorance. The possibility that science will soon advance to the point where all of us will have access to equivalently detailed foreknowledge of our own medical fates raises truly alarming ethical and political dilemmas. Such dilemmas form the subject of this book.

Notice that it is not just the DNA words themselves that can hurt. Knowledge about them can too—words in the other sense. We have been soothed and protected by ignorance. The principle of life insurance depends upon substantial—but crucially not total—ignorance of the future. Insurance works because actuarial calculations are no better than statistical approximations. In a world where everybody knew in advance the time and manner of their death, life insurance as we know it would break down. The only solution I can imagine for a humane society is a completely new system of universal insurance with forced equality of premiums, such as the 'People's Republic of Underwriters' proposed here by Justine Burley.

Such matters are difficult—which is why this book is important. We cannot simply invoke the right to privacy and claim that we patients, but not our insurers, should be allowed access to our genetic dossiers. As Bartha Maria Knoppers rightly points out, this would be unfair to insurers. If a genetic counsellor suddenly told you, and only you, that your life expectancy was dramatically shorter than the average for your age, mightn't your first act be to purchase life insurance on a massive scale—at premiums calculated on the false assumption that you are average? How could insurers be expected to tolerate this?

Similar problems arise from proposals to institute a national DNA database of all citizens.[1] The criminological benefits would be immense, but isn't it an intolerable infringement of personal liberty? Even if you retort (as I was at first tempted to) that only criminals have anything to fear, you will again be brought up short by the life insurance question. And even the most sanctimonious puritan might shrink from instituting a database that could disclose the true paternity of every child born in the land. Evidence suggests that about 10 per cent of putative 'fathers' in our society are not the biological fathers they, or their 'offspring', think they are. A strong case can be made that this particular species of ignorance is best not shattered.

Starting with Hilary Putman, several of our lecturers

[1]See Chapter 5, 'Barcodes at the Bar', of my book, *Unweaving the Rainbow*, Penguin Press, 1998.

appeal to Kant's moral principle that people should not be treated simply as means to the ends of other people. It undergirds one of the main arguments against human reproductive cloning. Giving parents a choice over the sex of a baby—a possibility briefly mentioned by Jonathan Glover—could similarly be condemned. Ironically, it is possible to make a good Kantian case in favour of this normally deplored practice. Given that people do, as a matter of fact, long for a son, or in other cases yearn for a daughter, the only recourse hitherto available to them is to keep trying until a baby of the desired sex turns up. As a result, we see many large families with sequences such as girl girl girl girl boy; or boy boy boy girl. Families are swollen by a long run of the less preferred sex before being rounded off by the first baby of the wanted sex. This incidentally has the effect of increasing population size, which some would consider undesirable in itself. It also falls foul of the Kantian maxim. Not only is the final son and heir (for example) being used as a means to the parents' ends—carrying on the family business, inheriting the title, or whatever it might be. So are at least some of the daughters, born not because they themselves were wanted (I hasten to add that they may still, of course, end up being deeply loved) but as a means to the end of producing a son. The delicate irony here is that the Kantian argument can be made to face both ways. At first sight, it obviously opposes allowing parents to choose the sex of their next baby. But I have now used it to make a case in favour of such choice, a case that might not easily have been anticipated. I note, without comment, an additional demographic consequence of allowing parental choice of sex. Those cultural and religious traditions that strongly favour boys over girls might be driven extinct through lack of women to bear children.

Several chapters in this book are stimulated, in their different ways, by Dolly the charismatic sheep, cloned from the nucleus of an adult who, for reasons that remain mysterious, is never dignified with a name. One definition of reproductive cloning is the creation of identical twins who are not the same age as each other. This definition is couched deliberately to reduce fear, and forestall that 'yuk' reaction to which several of these chapters call attention. Certainly, indentical twins could justly have felt slighted and hurt by some of the instant responses to Dolly's birth. Clones were said to lack 'individuality', 'separateness of will', and 'human dignity'—insults as offensive to identical twins as any slur meted out to victims of gross racial prejudice. Such 'nasal reasoning and olfactory moral philosophy'—John Harris's splendid phrase— has tainted many, from Presidents and Chief Rabbis down.

And yet, as several of these chapters persuasively argue, the difficulties raised by the prospect of human cloning are not to be lightly laughed off. John Harris rightly at this point invokes the libertarian principle that, if people really want something, we had better have pretty good reasons if we are going to forbid it: 'human liberty may not be abridged without due cause being shown.' Perhaps there are good reasons to forbid human cloning, and if there are I dare say they will be found in this book. Notwithstanding the well-made arguments of, for example, Hilary Putnam, Alan Colman and Ruth Deech, I remain persuaded by Harris's list of eight plausible situations in which human reproductive cloning could be justified on compassionate grounds.[2]

Other imaginable uses of cloning would be less condonable. Cloning oneself for reasons of personal vanity would fall foul of the Kantian exploitation principle. But (I was glad to see John Harris pointing out) so does engendering a child solely in order to produce a son and heir—a practice which, as we saw earlier, is widely indulged if not unreservedly approved. Or what about the fashionable mother who only ever relieves the nanny of her daughter in order to doll her up and parade her as a matching accessory? Don't many parents, even if only mildly, exploit their children as means to ignoble ends? And this says nothing about the environmental or educational dragooning of children in attempts (often vain, fortunately) to create cultural copies of their parents. It is widely accepted that parents' rights extend to educationally moulding their children to fit some expectation: Gordonstoun for outward bound toughening up, Eton, Inner City Comprehensive, Swiss Finishing School, or whatever it might be. More sinister, as Nicholas Humphrey argues in his lecture in the previous Amnesty series,[3] is that society's bending over backwards to respect religious tradition entails scant respect for the rights of children to opt out of, say ritual circumcision, or arranged marriage. Is

[2]My tolerant view of cloning is spelled out more fully in my essay in *Clones and Clones* (Ed. Martha C. Nussbaum and Cass R. Sunstein), Norton 1998.

[3]Nicholas Humphrey, 'What shall we tell the children?', Chapter 4 of *The Values of Science*, Oxford Amnesty Lectures 1997, Ed. Wes Williams, 1998.

cloning worse than all this simply because it involves *genetic* rather than environmental manipulation? Let me not pursue my own opinion. This is the kind of question the authors of this book are equipped to discuss. Going to the other extreme on the spectrum of children's rights, lawyers may warm to Hillel Steiner's startling notion that children might sue their parents for giving them the wrong genes.

Moralists are typically more sympathetic towards tissue culture cloning for medical purposes than they are towards reproductive cloning—making whole babies as clones of adults. Even here, words can be chosen to give tissue culture an unfortunate spin. Depending on how we choose our words, cultured tissues can be represented as 'belonging' to an unborn human victim. As Richard Gardner tells us,

monozygotic twins are more common during pregnancy than at birth. Thus competition between intimately associated fetuses must sometimes lead to the demise of one, not infrequently at so early a stage that no trace of it remains when the survivor is born.

Presumably the survivor benefits from the death of the competing twin. We have here a natural conflict which could be represented as hideously exploitative if it were humanly premeditated.[4] The storage of cloned tissue to provide spare parts for future transplants can be made

[4]For a fascinating account of Darwinian conflict in the uterus, including the surprising, yet compelling evidence of conflict between mother and fetus, see David Haig (1993), 'Genetic Conflicts in Human Pregnancy', *Quarterly Review of Biology,* **68**, 495–532.

to sound positively sacrilegious by people who see the tissue as essentially human, denied full personhood.

Such absolutist scruples can lead into a minefield of confusion. In my own contribution to the previous Amnesty series, I described a silly season stunt on British television by a well-known food connoisseur. A baby had been born, in happy circumstances, and the parents celebrated with a televised banquet in which they ate the placenta. The whole thing was portrayed as rather a lark. The presenter of the show, the famous food writer,

flash-fried strips of the placenta with shallots, and blended two thirds into a puree. The rest was flambeed in brandy and then sage and lime juice were added. The family of the baby concerned ate it, with 20 of their friends. The father thought it so delicious that he had 14 helpings. It tasted, he thought, like a Mediterranean beef dish, and the mother (she is only 20 and it is her first child) declared that serving the placenta would become a family tradition.[5]

Did the television company, did the presenter, did the parents or their twenty guests know, did they realize, that the placenta is actually a clone of the child itself? Its genotype is identical. You could think of the placenta as a discarded organ of the baby. Alternatively, it is as if there were identical twins, one of whom developed into a placenta, thereby sacrificing itself to the wellbeing of the other. Now read the culinary and gastronomic details again, and see how a few words of genetic knowledge can ruin the flavour. Not only might this public dinner be

[5]Richard Dawkins, 'The values of science and the science of values', Chapter 2 of *The Values of Science*, Oxford Amnesty Lectures 1997, Ed. Wes William, 1998.

indicted for the misdemeanour of exploiting a human clone. It could be said, by those who take an absolutist view of what it means to be human, to violate the last and deepest of our great taboos: the prohibition against cannibalism. Words again. So much mischief can lie in our choice of words.

If cannibalism is our greatest taboo, positive eugenics— the selective breeding of humans to favour particular mental or physical characteristics—is a candidate for the second greatest. With characteristic[6] intellectual courage, Jonathan Glover here takes on this thorny subject, rendered, as he points out, almost undiscussable by Hitler. Alan Ryan, in his engagingly written commentary on Glover's paper, reminds us that even the political left flirted with eugenics in the early part of the century, often using language that leaves us post-Hitlerians gasping.[7] Today the taboo is so strong and universal that it can warp our perception of reality. There are those on the left who, in their moral zeal against the merest suggestion of positive eugenics, leap to the conclusion that it is actually impossible in practice. This is wishful thinking on a grand scale. In non-human animals, positive eugenics can report dramatic successes: the prodigious milk yield of the Friesian cow, hens bred as living production lines to convert meal into eggs, wolves rendered into snuffling pekineses, fleet greyhounds or bear-like St Bernards—there is no reason to doubt that similar transformations could be wrought

[6]See, for example his *What Sort of People Should There Be?* (Penguin, 1984).
[7]See the quotations from H. G. Wells in my contribution to the previous Amnesty series, *loc. cit.*

from human raw material. Let us oppose this on moral grounds precisely because it *is* possible, not channel our moral indignation into futile denials of reality. Presumably none of the contributors to this book would disagree.

Similarly unrealistic denial bedevils the controversy over the very existence of genetic variation in human intelligence. There are those who, with the best moral and political motives, passionately want it not to be true that some people are genetically cleverer than others. Unfortunately, they are spitting into the wind. This can be demonstrated by simple Darwinian logic, without resorting to the complications of twin studies, and neatly bypassing the difficult decision whether this or that measure of IQ is a true measure of 'intelligence'. Simply define 'human intelligence' as that set of mental abilities which separates us from the other great apes and from the common ancestor that we share with them. Nobody denies that there is a large separation, and no well-informed person denies that it must have evolved by Darwinian selection. So, there exists some sensible meaning of brainy, according to which we are brainier than our ancestors, and this braininess has increased in evolution. Evolution cannot work unless there is genetic variation in the population. If there isn't, natural selection has nothing to work on. We have proved that, at least during the period of evolutionary swelling of the human brain, there was genetic variation in braininess between individual people. It is conceivable that, for some bizarre and unique reason, all such variation has recently disappeared. No such reason suggests itself, and I conclude that voguish tendencies to deny all genetic inequality in mental ability are rooted in cloud cuckooland, however laudable their political or moral motivation. Once again, let us take our moral stand in the light of unpalatable facts about the world, not found it on futile denials of their truth.

There are two further points that I, as an evolutionist, would diffidently add to the range of excellent and authoritative arguments aired in this book. The assumption that 'rights' automatically and self-evidently mean *human* rights is, as usual, strong. It is easy to sympathize with Ian Wilmut's need to distance himself from the tabooed prospect of human cloning. Yet the evolutionist, mindful of the cousinship of all mammals, cannot but wonder if the word 'obscene' isn't a little over the top, in the following quotation from the man who did, after all, preside over the making of Dolly:

it is surely obscene to even consider applying these techniques in humans for any reason.

Even the admirable Declaration of Human Duties espoused by Solomon Benatar mentions nonhuman life only in connection with protecting the environment from pollution and abuse, and upholding genetic diversity. Like most works of moral philosophy, this book largely ignores nonhuman suffering and any imaginable rights of individual nonhuman animals. Perhaps this is appropriate, for Amnesty International itself happens to concern itself—as it is entitled to—with that subset of putative rights that are human alone. Yet the evolutionist once again cannot help remembering that all animals are our cousins. But for the accidental extinction of the intermediates linking us to, for example, chimpanzees, we would be united to them by an interlocking chain of interbreeding: a daisy chain of the 'I've danced with a man, who's danced with a girl, who's danced with the Prince of Wales' variety.

If the intermediates had happened not to go extinct, our preoccupation with exclusively human rights could survive only on an intolerable platform of apartheid-like courts, sitting in odious conclave to decide whether particular apish intermediates would 'pass for human'. I am not here arguing for the extension of adult suffrage to our fellow African apes. Considerations such as these can just as plausibly lead us in the opposite direction: towards suspicion of the very idea of 'rights'; and towards even stronger suspicion of hard and fast lines wherever they may be drawn. I merely point out the evolutionary incoherence of absolutism. Any principle that depends on the accidental extinction of our close relatives cannot be all that fundamental or absolute.[8]

The other point I would make as an evolutionist is that the word 'Revolution', in the title of this book, is not lightly chosen. The genetic revolution may turn out more radical even than we ordinarily suppose. Evolution means change, by definition. But although forms and faunas change continuously over geological time, the process of evolution itself has been pretty much the same process,[9] at least since the rise of multicellular life just over half a billion years ago, and perhaps since the earlier origins of sex and the eucaryotic cell. Now our present era of artificial genetic revolution seems destined to change the very pattern of evolution itself. Never in the course of

---

[8] I developed this theme more fully in my contribution to *The Great Ape Project* (Ed. Paola Cavalieri and Peter Singer), Fourth Estate, 1993.
[9] *Unweaving the Rainbow*, Chapter 8.

evolutionary history has planned foresight played even the tiniest role. Elegant and apparently 'designed' as living adaptations undoubtedly are, their 'designs' are always for the short term benefit of individuals, as measured in units of immediate genetic success.[10] Humans find this stricture hard to credit, because our evolved brains are so good at forward planning. If short-term individual benefit is driving a species towards extinction in the long term, we find it hard to understand why natural selection does not step in to reverse the trend. Nevertheless it cannot, and it never has.[11]

Human interference at last puts paid to all that. For the first time, a mechanism exists whereby the future can be anticipated. For the first time, the course of evolution could be planned with the future in mind. This has actually been going on for some centuries in the form of eugenic selective breeding. Now, the genetic revolution offers the opportunity for humans to plan the mutational phase of the Darwinian process, as well as the selectional phase. Pigs might fly—if we plan the Darwinian manoeuvre sufficiently far ahead. Future historians may look back on the twenty-first century as a watershed in evolutionary history.

But that is a speculation for our distant posterity. This book is concerned with a nearer future, with ethical and moral issues that are going to be thrust earnestly in our faces during the next few decades. It is a most stimulating and thought-provoking book, not least because it venti-

[10]Richard Dawkins (1976), *The Selfish Gene*, Oxford University Press.
[11]George C. Williams (1966) *Adaptation and Natural Selection*, Princeton University Press, New Jersey.

lates opposing opinions, carefully thought out and elegantly expressed—and this includes chapters that I have not discussed, as well as those that I have. Moreover, these lecturers have not been shy to let their imaginations run free. All thinking people will find intellectual excitement in this book, and no person of goodwill can ignore the important questions vigorously and clearly debated here.

Contents

CONTENTS

Contributors

SOL BENATAR is Professor and Chairman at the University of Cape Town's Department of Internal Medicine and Chief Physician at Groote Shuur Hospital, where he also founded the UCT Bioethics Centre. Professor Benatar is a corresponding member of the US National Academy of Sciences' Committee on Human Rights, a member of the South African Medical Research Council's Ethics Committee, and Chairman of the UCT Research Ethics Committee. He has published extensively in the area of bioethics.

JUSTINE BURLEY is currently Simon Fellow in the Department of Government, University of Manchester, and a part-time Lecturer in Politics at Exeter College, Oxford. In addition to *The Genetic Revolution and Human Rights*, she is editor of *Ronald Dworkin and His Critics* and (with John Harris) *A Companion to Genethics*. Justine Burley is currently working on a monograph entitled *Genetic Justice*, forthcoming in the Oxford University Press Issues in Biomedical Ethics series.

ALAN COLMAN completed a Ph.D. in Molecular Biology in Cambridge under the supervision of Sir John Gurdon who pioneered the use of nuclear transfer to clone frogs from somatic cells. Dr Colman subsequently held academic posts at the University of Oxford (1974–6) where he was a demonstrator in reproductive physiology, the University of Warwick (1976–87) where he was made a

Senior Lecturer, and at the University of Birmingham (1987–93) where he was Professor of Biochemistry. He has worked full-time as Research Director of PPL Therapeutics Ltd. since 1993.

ROGER CRISP is Fellow and Tutor in Philosophy at St Anne's College, Oxford. He is the author of *Mill on Utilitarianism*, and edited the Oxford Philosophical Text of J. S. Mill's *Utilitarianism*. Dr Crisp is also editor of the journal *Utilitas*, and has also edited two collections on virtue ethics: *How Should One Live?* and (with Michael Slote), *Virtue Ethics*. He is a member of the *Analysis* Committee.

RICHARD DAWKINS is the Charles Simonyi Professor of the Public Understanding of Science at Oxford University, and a Fellow of New College. His field of expertise is evolutionary biology. His books are *The Selfish Gene*, *The Extended Phenotype*, *The Blind Watchmaker*, *River Out of Eden*, *Climbing Mount Improbable*, and *Unweaving the Rainbow*.

RUTH DEECH obtained a First Class Honours degree in Law from St Anne's College, Oxford (1965), and then an MA from Brandeis University. She was called to the Bar (Inner Temple) in 1967 and is now an Honorary Bencher of the Inner Temple. She became Fellow and Tutor in Law at St Anne's in 1970, and Principal in 1991. Ruth Deech is currently Chairman of the Human Embryology and Fertilisation Authority. She is also an Honorary Fellow of the Society of Advanced Legal Studies.

RICHARD GARDNER FRS obtained a First Class Honours degree in Physiology from Cambridge, and then a Ph.D.

on mammalian development under the supervision of Dr R. G. Edwards. He was appointed to a University Lectureship in the Department of Zoology, Oxford, in 1973, and was elected Student of Christ Church, Oxford, in 1974. He became Sir Henry Dale Research Professor of the Royal Society in 1978, and was elected to the Royal Society in 1979. Professor Gardner was Honorary Director of the Imperial Cancer Research Fund Developmental Biology Unit in the Department of Zoology, Oxford, from 1985 to 1996.

JONATHAN GLOVER became Professor of Ethics and Director of the Centre for Medical Law and Ethics at Kings College London in 1997. He was Chair of a European Commission working party and the author (with others) of its report *Fertility and the Family: The Glover Report on Reproductive Technology to the European Commission*. Professor Glover is the editor of *Philosophy of Mind* and (with Martha Nussbaum) *Women in Development*. He is also the author of *Responsibility*, *What Sort of People Should There Be?*, *Causing Death and Saving Lives*, and *I*. Professor Glover is currently working on a book about the Holocaust.

JOHN HARRIS is Sir David Alliance Professor of Bioethics at the University of Manchester, where he is also a Director of the Centre for Social Ethics and Policy and a Director of the Institute of Medicine, Law and Bioethics. In 1996, he was invited to join the new United Kingdom Government Advisory Committee on Gene Testing. His many publications include *Violence and Responsibility*, *The Value of Life*, *The Ethics of Human Biotechnology*, *Wonder Woman and Superman* (reissued with revisions as *Clones,*

Genes, and Immortality), and (with Justine Burley) *A Companion to Genethics*.

BARTHA MARIA KNOPPERS is Full Professor at the Faculty of Law, Universite de Montreal, Senior Researcher at the Centre for Public Law Research, and Counsel to the law firm McMaster Meighen. She is currently Chair of the Ethics Committee of the Human Genome Organization (HUGO) and was recently appointed to the Standing Committee on Ethics of the Medical Research Council of Canada (MRC). In 1996 she chaired the Organizing Committee of the First International Conference on DNA Sampling and Human Genetic Research: Ethical, Legal and Policy Aspects.

HILARY PUTNAM is currently Cogan University Professor at Harvard, where he has also been Walter Beverley Pearson Professor of Modern Mathematics and Mathematical Logic (1976–95), and Professor of Philosophy (1965–76). Professor Putnam has published widely in the areas of philosophy of mathematics, logic, mind, and language. His books include *Reason, Truth and History, Representation and Reality, Collected Papers* (vols. i, ii, iii), *Realism With a Human Face, Renewing Philosophy,* and *Words and Life.*

ALAN RYAN has been Warden of New College, Oxford, since 1996. In the previous thirty-five years he taught at many universities, including Princeton, Oxford, Essex, and the City University of New York. He has written extensively on issues in applied ethics, and his most recent book (1998) is *Liberal Anxieties and Liberal Education.*

HILLEL STEINER is Professor of Political Philosophy in

the Department of Government and in the School of Philosophical Studies at the University of Manchester. His recent book, *An Essay on Rights* (awarded the W. J. M. Mackenzie Prize), advances a novel theory of justice that is focused on the coherent integration of individuals' entitlements to personal liberty with those to equal treatment. His current work is concentrated primarily on international and genetic justice.

IAN WILMUT, a scientist of international renown, is currently Principal Investigator in the Division of Development and Reproduction at the Roslin Institute, Edinburgh. At the Institute he is contributor/joint leader of the team that produced transgenic sheep which secrete large quantities of human proteins in their milk, work that is now being commercialized. The focus of his research over the past three years has been on the factors regulating embryo development in sheep after nuclear transfer. This research led to the birth of 'Dolly' the sheep, the first mammal to be 'cloned' using adult cells.

JONATHAN WOLFF is Head of Department and Reader in Philosophy at University College London. He is the author of *Robert Nozick: Property, Justice and the Minimal State* and *An Introduction to Political Philosophy*, and editor of the *Proceedings of the Aristotelian Society*. He has published papers in contemporary political philosophy and the history of political thought, and is currently working on the idea of an egalitarian social ethos.

Acknowledgements

It is not common practice to thank the contributors to a book one is editing. However, it seems more than appropriate to do so in the case of this volume given its particular origin and purpose. *The Genetic Revolution and Human Rights* comprises the 1998 Oxford Amnesty Lectures, the proceeds from which are donated to the research section of Amnesty International. I am especially grateful to Sol Benatar, Jonathan Glover, John Harris, Bartha Maria Knoppers, Hilary Putnam, Hillel Steiner, and Ian Wilmut, the lecturers and thence main contributors to this collection of essays, for devoting their time and expertise to the series. Roger Crisp, Ruth Deech, Richard Gardner, and Alan Ryan also worked hard for the 1998 Oxford Amnesty Lectures, each introducing and providing a discussion article for a lecture. G. A. Cohen, Michael Dummett, and Liz Frazer respectively, graciously introduced a speaker; Jo Wolff provided an insightful commentary on Hillel Steiner's talk and Alan Colman provided one on Hilary Putnam's talk; Richard Dawkins, who has been a tireless supporter of the Oxford Amnesty Lectures, kindly agreed to write the Foreword to this volume. Without the participation of the aforementioned individuals, neither the lectures nor this volume would have been possible.

Chris Miller, Nick Owen, Fabienne Pagnier, Steve Shute, Deana Rankin, and Wes Williams, Directors of Oxford Amnesty Lectures Ltd., merit warm thanks for their assistance with the organization of this year's series.

Sue Waldman, of the Sheldonian Theatre (once again) proved invaluable to the success of the lectures, as did Blackwells, the city and university branches of Amnesty International, and staff members of Exeter, New, and the Queen's Colleges, Oxford, for making the lecturers welcome when in residence. Many more dedicated people helped behind the scenes with leaflet distribution, ticket sales, ushering, and so on.

The Times Higher Educational Supplement has generously sponsored the Oxford Amnesty Lectures for many years and deserves special mention for so doing.

I am grateful to Angus Phillips of Oxford University Press for his excellent guidance, and enthusiasm for this book.

I should also like to acknowledge the support of the Institute of Medicine, Law and Bioethics of the University of Manchester.

Finally, along with the other Directors of Oxford Amnesty Lectures Ltd., I wish to express appreciation to members of the general public and of the academic community in and around Oxford whose interest in rights-related issues makes the lectures worthwhile.

Cloning People

Hilary Putnam

When Ian Wilmut and his co-workers at the Roslin Institute in Edinburgh announced early in 1997 that a sheep
'Dolly' had been successfully cloned, there was an amazing
spontaneous reaction. People all around the world felt that
something morally problematic was threatening to happen.
I say 'threatening to happen', because most of the concern
centres not on the cloning of sheep, guinea pigs, etc., but
on the likelihood that sooner or later someone will clone
a human being. And I say 'spontaneous reaction' because
this is probably not a case in which a moral principle that
had been formulated either by traditional ethical sources,
religious or secular, was clearly violated or would be violated if we succeeded in cloning people, or someone would
be deprived (if the feared possibility materializes) of what
is already recognized to be a human right. Of course,
some proposed uses of cloning do violate the great Kantian
maxim against treating another person solely as a means—
for example, cloning a human being solely so that the
clone could be a kidney donor or a bone-marrow donor;
but these are probably not the cases that came immediately
to people's minds.[1] I will argue that cloning humans (if
and when that happens) may, indeed, violate human
dignity even when the purpose is not as blatantly instrumental as producing an organ donor; but I don't think

the spontaneous reaction I described resulted from a considered view as to how this would be the case.

The reaction was the sort of reaction that gives rise to a morality rather than the product of a worked through reflection.[2] Such reflections can be the source of moral insight, but they can also lead to disastrous moral error. I do not conclude from the fact that a ban on, or restriction of, or condemnation of the practice is not easily derived from already codified moral doctrines that it should be presumed that cloning presents no moral problems; on the contrary, I shall argue that it poses extremely grave problems. What I want to do is say why the issue is a grave one—for I believe that the spontaneous reaction is justified—and to begin the kind of reflection that I believe we need to engage in, the kind of dialogue that we need to have, to make clearer to ourselves just what issues are at stake.

The scenario with which I shall be concerned—concerned with because it is, I believe, precisely the sequence of events that people fear may transpire—is the scenario in which (1) we learn how to clone people; and (2) the 'technology' becomes widely employed, not just by infertile couples, but by ordinary fertile people who simply wish to have a child 'just like'[3] so-and-so. Why do so many of us view this scenario with horror?

Although this will be my central question, it must be mentioned that, quite apart from the spontaneous horror I described, there are additional grounds, including some obvious utilitarian grounds, for being worried about the possible misuse of cloning technology. (And anyone who relies on 'market mechanisms' or 'consumer sovereignty' by themselves to prevent misuse of anything must have his head in the sand.) For one thing, even techniques of 'bioengineering' that seem utterly benign, such as the techniques that have so spectacularly increased the yields of certain crops, have the side-effect, when used as widely as they are being used now, of drastically reducing the genetic diversity of our food grains, and thus increasing the probability of a disaster of global proportions should a disease strike these 'high-yield' crops. If cloning should be used not just to produce animals for the production of medically useful drugs, but to produce 'twins' of, say, some sheep with especially fine wool, or some cow with especially high milk-yield or beef, and the practice should catch on in a big way, the resulting loss of genetic diversity might well be even more serious. And if cloning ever became a really popular way of having babies, then the question of the result of the practice on human genetic diversity would also have to be considered. But, although these issues are obvious and serious, it is not my purpose to address them further at this time, for I do not believe that *they* are what lies behind the spontaneous reaction to which I alluded.

The Family as a Moral Image

When I say that I want to engage in a reflection that will help us become clearer about the cloning issue, I do not mean that I want in this chapter to propose a specific set of principles or a methodology for deciding whether cloning people could ever be justified (or deciding when it would be justified). In a book I published a decade ago,[4] I argued that moral philosophers who confine themselves to talk about rights, or virtues, or duties, are making a mistake:

that what we need first and foremost is a moral image of the world. A moral image, in the sense in which I use the term, is not a declaration that this or that is a virtue or a right; it is rather a picture of how our virtues and ideals hang together with one another, and what they have to do with the position we are in. I illustrated what I meant by the notion of a moral image by showing that we can get a richer appreciation of the Kantian project in ethics if we see the detailed principles that Kant argued for as flowing from such a moral image (one that I find extremely attractive): an image of human equality. I claimed that the respect in which human beings are equal, according to Kant, is that we have to think for ourselves concerning the question of how one ought to live, and that we have to do this without knowing of a human *telos*, or having a clear notion of human flourishing. I argued that this notion of equality, unlike earlier ones, is incompatible with totalitarianism and authoritarianism. But although I find this image very appealing, I argued that it is not sufficient, and I spoke of us needing a plurality of moral images. Commenting on this, Ruth Anna Putnam has written:

I think that this reference to pluralism must be understood in a twofold way. On the one hand, the image of autonomous choosers is too sparse, it needs to be filled out in various ways, we need a richly textured yet coherent image. On the other hand, we need to recognize that people with different moral images may lead equally good lives. I do not say, of course, that all moral images are equally good, there are quite atrocious moral images: I am saying that there are alternative moral images by which people have led good lives, and that we can learn from their images as they can learn from ours. One of our very deep failings is, I suspect, a tendency to have moral images that are too sparse.[5]

One more remark about moral images before I return to our primary topic. In stressing the importance of moral images I do not mean to suggest that in some way moral images are a foundation from which moral principles, lists of rights and virtues, etc., are to be derived. I have never been a foundationalist, in ethics or in philosophy of science or in any other area of philosophy. One can raise the question of justification with respect to a moral image of the world (or a part or aspect of the world or of life) just as we can raise it about any particular and partial value.[6] As I put it in *The Many Faces of Realism*: 'the notion of a value, the notion of a moral image, the notion of a standard, the notion of a need, are so intertwined that none of them can provide a "foundation" for ethics.'[7] Thus Ruth Anna Putnam has observed:

It follows, it seems to me, that when he says that a moral philosophy requires a moral image he must not be understood to say that a moral philosophy requires a foundation, or an ultimate justification for whatever values it espouses, a foundation that can only be provided by a moral image. Rather we must understand him to say, and with this I agree, that without a moral image any moral philosophy is incomplete.[8]

What I want to do now is to describe some moral images of the family, moral images which turn out to influence how we think not just about the family but about communal life in general.

That we do use images derived from family life in structuring our whole way of looking at society and our whole way of seeing our moral responsibilities to one another is not difficult to see. I became keenly aware of this many years ago as the result of a conversation with my now late mother-in-law, Marie Hall. In her youth, Marie had lived a committed and dangerous life, being active in the anti-Hitler underground in Nazi Germany for two years before escaping the country and eventually making her way to the United States, and she continued to show an admirable commitment to a host of good causes until she died (she lived to be 86). I admired her and loved her and adored her not just for her commitment but also for the vitality and humour that accompanied everything she did, and I constantly 'pumped' her about her attitudes and activities during various periods in her life. On one occasion I asked her why she was inspired to make such efforts and run such risks for a better world, in spite of all the setbacks. I was amazed when she answered by saying simply 'all men are brothers'. She meant it. For her, 'all men are brothers' was not a cliche; it was an image that informed and inspired her whole life. It was at that moment that I understood the role that a moral image can play.

The particular image that inspired Marie Hall, the image of us as all brothers (and sisters), played a huge role in the French Revolution and after. The great slogan 'Liberty, Equality, Fraternity' listed, perhaps for the first time, fraternity as an ideal on a par with equality and liberty. And to this day union members and members of oppressed groups frequently refer to one another as brothers and sisters. Of course, we all know that 'in real life' siblings frequently do not get along. Yet we do have images of what ideal family relations should be, and it is clear that these images are enormously powerful and can move large numbers of people to do both wonderful and (as the French Revolution also illustrates) terrible things.

The use of the image of an ideal family as a metaphor for what society should be is not confined to the West. Confucian thought, for example, has an elaborate picture of an ideal harmonious (and also hierarchically ordered) family which it consciously uses as a guiding metaphor in thinking about what an ideal society would be. What I want to turn to now is the following questions: what should our image of an ideal family be, and what bearing does that have on whether we do or don't view the 'cloning scenario' with horror?

Moral Images of the Family

I began this chapter by describing a 'scenario' that is, I claimed, evoked by the prospect of cloning people. In that scenario, first, we learn how to clone people: and secondly, the 'technology' becomes widely employed, not only by infertile couples, but by ordinary fertile people who simply wish to have a child 'just like' so-and-so. I want now to see how our evaluation of this scenario will be evaluated from the standpoint of different possible (or immoral) images of the family.

Let us begin with an image that we would not regard as a moral image at all. Imagine that one's children come to be viewed simply as parts of one's 'lifestyle'. Just as one has the right to choose one's furniture to express one's personality, or to suit one's personal predilections, or (even if one does not wish to admit it) 'to keep up with the Joneses', so, let us imagine that it becomes the accepted pattern of thought and behaviour to 'choose' one's children (by choosing whom one will 'clone' them from, from among the available relatives, or friends, or, if one has lots of money, persons who are willing to be cloned for cash). In the Brave New World I am asking us to imagine, one can have, so to speak, 'designer children' as well as 'designer clothes'. Every narcissistic motive is allowed free reign.

What horrifies us about this scenario is that, in it, one's children are viewed simply as objects, as if they were commodities like a television set or a new carpet. Here what I referred to as the Kantian maxim against treating another person only as a means is clearly violated.

In a recent article,[9] Richard Lewontin, a great evolutionary biologist and an outspoken radical, argued that it is hypocritical to worry about this as long as we allow capitalist production relations to exist.

We would all agree that it is morally repugnant to use human beings as mere instruments of our deliberate ends. Or would we? That's what I do when I call in the plumber. The very words 'employment' and 'employee' are descriptions of an objectified relationship in which human beings are 'things to be valued according to externally imposed standards'.[10]

But Lewontin is confused as to what the Kantian maxim means. Even when someone is one's employee, there is a difference between treating that someone as a mere thing and recognizing his or her humanity, That is why there are criteria of civilized behaviour with respect to employees but not with respect to screwdrivers. An excellent discussion of what these criteria require of one and how they are related to the Kantian principle can be found in Agnes Heller's *A Philosophy of Morals*.[11]

Let me now describe the moral image I want to recommend, and then consider some alternatives which are not as blatantly narcissistic as the one in which children are treated simply as adjuncts to one's so-called lifestyle. First of all, if our image is to be a moral image at all, it should conform to the Kantian maxim. In an ideal family, the members regard one another as 'ends in their own right', as human beings whose projects and whose happiness are important in themselves, and not simply as they conduce to the satisfaction of the parents' (or anyone else's) goals. Moreover, it should be inspired by the Kantian moral image I described at the outset, the image which assigns inestimable value to our capacity to *think for ourselves in moral matters*. Here, perhaps, one should think not only of Kant but also of Hegel, who, in *The Philosophy of Right*, argues that the task of good parents is precisely to *prepare children for autonomy*. The good parent, in this image, looks forward to having children who will live independently of the parents, not just in a physical or an economic sense, but in the sense of thinking for themselves, even if that means that they will inevitably disagree with the parent on some matters, and may disagree on matters that both parents and child regard as important. Rather than regarding the tendency of one's children to think for themselves, to disagree with one when they reach the age of reflection, to have tastes and projects one would not have chosen for them, one should welcome each of these. I note that although the Confucian image of the good family to which I referred earlier would agree that all the members of a good family should value one another as ends and not as mere means, the valorization of autonomy is foreign to classical Confucianism, with one exception:

having the capacity to stand up for the right, when the alternative is clearly evil, is valued in the Confucian tradition. What is missing is the value of independent thinking about what morality requires of one. (Contemporary NeoConfucian thinkers are struggling with the problem of incorporating such Enlightenment values as autonomy and equality (including gender equality as well as equality between younger and older siblings) into a broadly Confucian framework, with interesting results.)

If one does accept the values I have described, and incorporates them into one's moral image of the family, then, I think, there is one further value that is important (and very natural) to add, and that is the value of willingness to accept diversity. As things stand now (I speak as a parent of four children and a grandparent as well), the amazing thing about one's children is that they come into one's life as different—very different—people seemingly from the moment of birth. In any other relationship, one can choose to some extent the traits of one's associates, but with one's children (and one's parents) one can only accept what God gives one to accept. And, paradoxically, that is one of the most valuable things about the love between parent and child: that, at its best, it involves the capacity to love what is very different from one's self. Of course, the love of a spouse or partner also involves that capacity, but in that case loving someone with those differences from oneself is subject to choice; one has no choice in the case of one's children.

But why should we value diversity in this way? One important reason, I believe, is precisely that our moral image of a good family strongly conditions our moral image of a good society. Consider the Nazi posters showing 'good' Nazi families. Every single individual, adult or child, male or female, is blond; no one is too fat or too thin, all the males are muscular, etc.! The refusal to tolerate ethnic diversity in the society is reflected in the image of the family as utterly homogeneous in these ways.

I am not claiming that a positive valuation of diversity *follows from* a positive valuation of autonomy. On the contrary. There have been societies which valued autonomy and moral independence while devaluing diversity to the extent of believing in their own 'racial superiority' and engaging in widespread sterilization of those who were seen as 'unfit'. I am thinking of the social-democratic Scandinavian countries which passed sterilization laws in the 1930s 'with hardly any secular or religious protest'.[12] Here is Daniel Kevles's description: 'Eugenics doctrines were articulated by physicians, mental-health professionals and scientists, notably biologists who were pursuing the new discipline of genetics. They were widely popularized in books, lectures, articles to the educated public of the day, and were bolstered by the research that poured out of institutes for the study of eugenics or "race biology" that were established in a number of countries, including Denmark, Sweden and the United States.' The experts raised the spectre of social 'degeneration' insisting that 'feeble-minded' people—to use the broad-brush term then commonly applied to persons believed to be mentally retarded—were responsible for a wide range of social problems and were proliferating at a rate that threatened social resources and stability. Feeble-minded women were held to be driven by a heedless sexuality, the product of biologically grounded flaws in their moral character that led them to prostitution and illegitimacy. Such biological analyses of social behaviour found a receptive audience among middle-class men and women, many of whom were sexually prudish and apprehensive about the discordant trends of modern urban, industrial society, including the growing demands for women's rights and sexual tolerance.

The Scandinavian region's population was relatively homogenous, predominately Lutheran in religion, and Nordic in what it took to be its racial identity. In this era, differences of ethnicity or nationality were often classified as racial distinctions. Swedish analysts found that the racial purity of their country might eventually be undermined, if only because so many Nordics were emigrating, Swedish speakers in Finland feared the proliferation of Finnish speakers, holding them to be fundamentally Mongols and as such a threat to national quality.[13]

Our moral image of the family should reflect our tolerant and pluralistic values, not our narcissistic and xenophobic ones. And that means that we should welcome rather than deplore the fact that our children are not us and not designed by us, but radically Other.[14]

Am I suggesting, then, that moral images of the family which depict the members of the ideal family as all alike, either physically or spiritually, may lead to the abominations that the eugenics movement contributed to? The answer is, 'very easily'. But my reasons for recommending an image of the family which rejects the whole idea of trying to pre-design one's offspring, by cloning or otherwise, are not, in the main, consequentialist ones. What I have been claiming is that the unpredictability and diversity of our progeny is an intrinsic value and that a moral image of the family that reflects it coheres with the moral images of society that underlay our democratic aspirations. Marie Hall was willing to risk her very life in Hitler's Germany for the principle that 'all men are brothers'. She did not mean that 'all men are identical twins'.

In closing, since this is a lecture devoted to human rights, I suppose that I should mention human rights at least once. If 'rights' talk has not figured in my discussion, it is because, as explained, I believe that our conception of rights, values, duties, etc., needs to cohere with a moral image which is capable of inspiring us. But perhaps one novel human right is suggested by the present discussion: the 'right' of each newborn child to be a complete surprise to its parents!

Why Human Cloning Should not be Attempted

Alan Colman

On 27 February 1997 my colleagues at PPL Therapeutics, Angelika Schneike and Alex Kind, along with Ian Wilmut, Keith Campbell, and Jim McWhir of the Roslin Institute, announced in the journal *Nature* the arrival of Dolly, the first mammal cloned from an adult cell. The reaction of the scientific community was on the whole friendly, if somewhat incredulous. In contrast, the reaction of the public at large was overwhelmingly negative, primed to some extent by a media weaned on a diet of previous cloning scare stories and pulp fiction. The issue of course was not that this new technology created an abomination, a freak animal—Dolly was, and remains, a rather handsome sheep—it was the fear that the techniques could, and some said would, be applied to humans.

I think it clear enough that the research that led to Dolly, if continued in animal models, will be of enormous benefit to human beings. But I agree with Hilary Putnam that cloning should not be attempted in humans. Below I shall argue, as a further comment on Putnam's position, that human cloning—which, for the purposes of what follows, should be read as *reproductive* cloning—is unethical for two main reasons: it would be unsafe and it would be inefficient.

The creation of Dolly was a major scientific achievement which began with the transfer of the nucleus of an adult cell to an unfertilized egg taken from a donor animal by a process known as cell fusion. The so-called reconstructed embryo is cultured and eventually returned to the womb of a foster mother and brought to term. It is essential to stress that from over 430 attempted fusions, 277 reconstructed embryos were made in this way. Of these, only 29 survived to the stage that they could be returned to 13 foster mothers and only one survived to term. In other experiments, some of the lost fetuses were abnormal (some severely so), and were aborted. In addition, fetal abnormality did not always become manifest before an advanced stage of development. Thus, as things stand now in animal models, the technique of nuclear transfer has a low success rate. What implications does this have for human cloning?

We have grounds to suppose that attempting cloning in humans would be dangerous on numerous counts. Think of the waste of human material. How many embryos would have to be discarded or would be lost? Think of the suffering, both physical and psychological, of the biological parents and/or surrogate mother. How many surrogates would be needed? How many times would fertilized eggs have to be implanted into a woman's womb before one implantation was successful? How many fetuses might be miscarried or aborted? Think of any cloned child who was seriously deformed as a product of the cloning procedure.

At this juncture in my argument, advocates of human cloning will say that all the uncertainties could eventually be dispelled by first perfecting the technique in animal models. I disagree for the reason that there are too many differences in the reproductive physiology and embryology between different mammalian species, including humans. We simply could never be confident that high success rates in one species, if achieved, could be replicated in humans without a lot of human experimentation. Of course it is true that certain currently available assisted conception techniques, such as *in vitro* fertilization, have very low success rates. However, apart from the technical challenges implicit in the extension of cloning to humans, there are other, more subtle concerns.

There is evidence to suggest that as we get older, our cellular DNA (the building-blocks of our genes) gradually accumulates mutations and suffers other changes which account for the fact that we are increasingly liable to develop cancer as we get older, and might also explain how our cells, and therefore our bodies, are affected by ageing. Thus a human cloned from an adult cell, as was Dolly, might have a higher risk of cancer or experience premature ageing. And, importantly, this might not be known for some years. Moreover, it is worth labouring the point that even if Dolly, for example, ages 'normally', this does not mean that a cloned human being would.

Neither doctors nor society should condone a practice which carries the sorts of high risks identified above. Were they to, human cloning would quite likely join thalidomide in the teratogenic hall of infamy.

It might be objected here that all new medical advances are potentially unsafe, and no progress would be made if safety was the sole consideration. However, risk-to-benefit ratios must always be considered before new treatments are sanctioned. The application of such ratios to new reproductive treatments are particularly problematic given that we must consider risks to the egg donor, the womb donor, and the unborn child. Even if the egg donor and surrogate gave their informed consent, it remains the case that creating a new life by nuclear substitution would be likely to result in the conception of a fetus that suffered abnormality caused by the actual procedure of transferring genetic material from adult cell to gamete, or that the technique would cause a person to suffer a genetically related disorder later in life. Such occurrences would be most unfortunate not least because they could have been avoided.

There may be scope, however, to apply cloning tech-nology in humans in a way that would avoid some of the aforementioned experimental uncertainties altogether, and also limit others. I have in mind here the case of a prospec-tive mother who has a genetic disease which is not attributable to the main body of genes found in the nucleus, but to genes elsewhere in the cell (the mitochondria). With all existing methods of conception, both natural and assisted, all the children of such women would inherit the disease-causing genes. It has been sug-gested that women who possess defective mitochondria could conceive normally and then a cell from the doomed embryo could be fused to the unfertilized egg supplied by a 'healthy' human egg donor. If successful, this would result in a child free from the disease which has a unique genetic blueprint and one made up from equal contri-butions from the original couple (apart from the offending non-nuclear genes). Additionally, the fact that such an early donor cell would be used might avoid the risk of accumulated DNA damage which, as mentioned above, could affect use of adult cells.

But, with a frequency of mitochondrial disease of 1 in 20,000, I judge that even in this sort of case the procedural risk greatly outweighs the benefit. While the circumscribed use of cloning technology just outlined would be less prone to experimental uncertainties, the cell 'surgery' involved would carry risks which could not be justified quantitatively by the benefit.

I therefore conclude that human reproductive cloning, in all its guises, is unethical. I have argued elsewhere that a child, so manufactured, and that I believe is an appropriate term, would be a twenty-first-century circus act.[1] And so, along with Hilary Putnam, I think that human cloning would be deleterious to any cloned child who would lack its own unique identity. Even if the child's uniqueness is not compromised, nuclear transfer in humans, as I have argued, is patently unsafe and inefficient; the risks greatly outweigh any marginal benefit.

People may, then, be forgiven for asking whether nuclear transfer research in animals should ever should have been done and whether it should continue. The answer has to be an unequivocal yes. As Ian Wilmut has spelled out (see Chapter 2), the potential for non-human cloning work to provide helpful solutions to many problems from which humans suffer is vast.

Dolly: The Age of Biological Control

Ian Wilmut

In this chapter I wish to present the reader with several exciting possible applications of nuclear transfer technology, or cloning technology as it is popularly called. Nuclear transfer is the technique by which Dolly was created.[1] What makes Dolly so unique is that she is the first mammal ever to be cloned using the nucleus of an already differentiated cell. In all of the furore that followed our announcement of Dolly's birth at the Roslin Institute, and of her subsequent pregnancy, there has been relatively little attention paid to how revolutionary this step in nuclear transfer technology in fact was. What Dolly shows is that a nucleus from an adult animal cell can be reprogrammed to become totipotent; it can, in other words, be altered to allow the full range of gene expression required to create a complete animal. We have thus entered what I have termed in the title an age of biological control, an era in which we will progressively become able to prevent as well as treat human diseases which currently cause terrible suffering, both by better understanding the molecular structure of cell development and modifying it. I am hopeful that with time and more information people will increasingly come to share my view of the potentially enormous benefits to human beings that further advances in nuclear transfer technology might bring. Before I discuss these I wish first to stress that this technique is very much in its infancy.

Dolly was the result of 277 attempts to fuse an adult nucleus with an egg. Of these 277, twenty-seven embryos developed normally for the first week, but only one developed to term. Analysis of the results with a variety of different cell types showed that 50 per cent of the fetuses were lost in the last two-thirds of development, as compared with a rate of 6 per cent loss during that period in natural procreation. In addition, 20 per cent of the lambs born live died shortly after birth. Some of the sheep fetuses were unusually large: the development of their heart, liver, and kidneys was profoundly changed. These abnormalities occurred because the clock in the nucleus was not reset properly. Although we do not yet understand the precise mechanisms behind such changes in development, I am confident that one day we will, and also that we will learn either to manage or eliminate them. My optimism is fuelled by the recent birth of Cumulina the mouse, and her twenty-one sibling clones.[2] There are also reports of calves being born in Japan after nuclear transfer from a cell taken from an adult. In total, then, offspring have been produced from three mammalian species by fusing adult nuclei with eggs. And so I predict that it will be possible some time in the future to master this technique, marshal its widespread use, and better understand the implications of the complexities involved for human well-being. As I see it, if we do succeed in perfecting nuclear transfer technology, there are three main ways in which it could be of immense value to us: (1) making copies of the same animal; (2) making precise changes in DNA; and (3) making cells to order.

Copying

Why should we want to make copies of the same animal?

First, duplicating animals is very helpful for drug trials. If we make exact genetic copies of an animal we can test the efficacy of a drug with much greater accuracy and, at the same time, pinpoint the way(s) in which it works. It bears noting that the possibility of making genetically identical animals should limit, at least to some extent, the amount of animal experimentation that is carried out— fewer animals need be experimented on if the influences of differences between animal subjects in experiments are minimized.

Second, if genetic replicas of farm animals are made, both animal health and agricultural output could be much enhanced. Only genetically robust stock would be produced; healthy, productive animals, disease-free cows with high milk yields, would benefit both farmers and consumers. And we could always be sure of producing animals of the required sex, cows rather than bulls, for example, when the objective was to produce milk, but bulls when beef animals were required. As male animals are more efficient at producing lean meat there would be an advantage in making copies of bulls with good carcass characteristics.

Of course, the technology might be used to copy humans. For the most part this is a patently unattractive possibility, as far as I am concerned. At present, with the likelihood of abortions and infant deaths, it is surely obscene to even consider applying these techniques in humans for any reason. The methods will be improved, but it is not possible to predict when or to what extent they will be improved. However, in addition to this question of safety there are ethical concerns over potential uses of nuclear transfer to copy existing people. The people who have approached me about cloning have given three main reasons for wanting to clone a human: to bring back a lost relative, selective breeding, and to treat infertility.

A little while ago I received a telephone call from a woman who had lost her 2-year-old to leukaemia the previous Sunday. Could I bring her child back? Tragic. I suspect that most of us, if we could do this, would do it. We would wave the magic wand and summon the child back. But, in fact, what would happen were you to make a copy of someone is that you would make a new, different individual who might very well grow in a different way and become a different sort of person. Incidentally, all our ideas that we might have been different, stories that we tell ourselves about the crucial turning-points in our lives, involve a similar, complex understanding: 'I could have been a contender' does not mean that I could have had different genes. Our personality is apparently determined about half by our genes and half by what happens to us.

Selective breeding I find extremely difficult to fathom. Indeed, recently, while in California, I was shocked to remember that there are human sperm banks. The use of cloning for selective breeding also shocks me and why someone should attempt it for this purpose is puzzling. Great emphasis is placed, particularly in the United States, on reproductive choice. But surely this freedom must be accompanied by responsibility. My gravest concern is for the child. Making a copy is not treating the child in the way that he or she deserves, not treating the child as an individual.

Using cloning to treat infertility raises first and foremost in my mind concern about family relationships. If my wife and I are infertile, and we decided that I should be cloned, could I have an effective, healthy relationship with someone who is a copy of me? Could my wife? And, importantly, could the child have a good relationship with me? Although I think it eminently possible for one's attitude towards an adopted child to be the same as it is to one's own offspring, I strongly doubt that the same parity of attitude could be achieved in a family in which there was a genetic replica of one of the parents.

But there is one way nuclear transfer technology might be used in procreation that I do find attractive. This is its use to replace the mitochondrial DNA in an egg. Mitochondria are the small bodies in each of our cells which supply energy. They contain DNA, which is subject to error (mutation) leading to diseases in just the same way that chromosomal mutation may cause disease. However, in the case of mitochondria we inherit those only from our mothers. A woman suffering from mitochondrial disease knows that her children will inherit the same condition. In principle, there is no reason why the embryo nucleus could not be removed from the defective egg and be placed in a recipient egg cell, itself enucleated. The recipient egg would be provided by a woman known not to have similar damage to her mitochondria, with her full informed consent. The resulting child would be exactly as it would have developed, except that it would not suffer the disease associated with mitochondria. The catch with this is that it would be possible to make multiple copies of the embryo—you might start with a thirty-cell embryo and end up with six fertilized eggs. Done thoughtfully, however, this method of nuclear transfer could provide a way to treat currently untreatable mitochondrially carried diseases.

Making Precise Changes

The second application of nuclear transfer technology is its use in making precise changes in the DNA of animals. Indeed, this was the original, and continues to be the main, aim of nuclear transfer research. The principal idea is that once a gene has been discovered we can study its effects in an animal grown from a suitably prepared embryo. The diabetes research conducted in the late nineteenth and in the early twentieth centuries serves as a good analogy here. Frederick Banting and Charles Best discovered insulin after the removal of a dog's pancreas caused the symptoms of diabetes. Already we can do the genetic equivalent to taking out the pancreas, by preventing a gene from functioning. In one ongoing project, for example, mutation associated with human cystic fibrosis has been introduced into mice. Cystic fibrosis is a disease caused by errors in a single gene which encodes the production of a protein on the cell surface. Patients suffer congestion of the lungs and are vulnerable to pneumonia. The disease in mice is different to that in humans and it has been suggested that it might be better studied in sheep. This would mean that diseased sheep would be produced by introducing the mutation into sheep to create multiple subjects for the study of the pathology of cystic fibrosis under more highly controlled conditions than is presently feasible. It would be much easier to study the disease in larger animals and the sheep has lungs and certain genes which closely resemble those in humans.

A second way in which nuclear transfer technology could be used to effect precise changes in DNA is in the manufacture of certain human proteins in the milk of animals such as sheep and cows. For example, the human gene for clotting factor IX (which is required by haemophiliacs) can now be transferred into sheep. Soon we should be able, using nuclear transfer technology, to produce large numbers of identical therapeutic milk producers and hence all the clotting factor IX required.

In addition, nuclear transfer technology could be used to create transgenic animals for xenotransplantation. At present, over 160,000 patients die each year before a suitable organ becomes available from a human donor. If an organ is transferred from an animal, the organ is destroyed rapidly by an immune response. The idea of genetically modifying porcine tissues, for example, to resist human immune responses is gradually gaining more support. Such animals could be produced en masse to act as a source of life-saving vital organs for humans suffering renal, liver, or heart failure. This is an attractive application of the technology when one considers that so many people die each year for want of transplant organs.

The first and the last of the three examples just mentioned pose palpable ethical dilemmas. In the first case one is deliberately diseasing an animal, and in the last case one invites the possibility of introducing an entirely new disease into the human population. My own view with respect to the first dilemma is that so long as the animal receives the same treatment as would a human patient, and there is a good chance that the experiment will succeed, it should be performed. As for the second, opinion remains divided over the precise risks to humans of xenotransplantation. The chief worry of opponents is that animal-to-human transplants would introduce viral infections from animals. New diseases might, it is feared, enter the human gene pool and wreak misery.

Finally, it may well be possible one day to use the technique of nuclear transfer to prevent known genetic diseases by correcting the defect in genes. There has been a good deal of talk about 'designer babies' in the press. Less frequently discussed is the possibility of genetically engineering corrections to defective genes such as the one for cystic fibrosis. Precise changes in DNA of this kind might also enable us to create resistance to malaria (a big killer in the developing world), prevent baldness, or enhance athleticism. The idea of correcting for overt disease in this way involves pragmatic judgements about how much we really know about genes. I am not confident that we do know enough at present, but it is likely that within the lifetimes of the younger members of our society germ-line gene therapy will be used in humans.

Producing Cells for Further Use

The third and final main possible application of nuclear transfer technology is its use in making human cells to order. By this I mean that we may one day be able to take differentiated cells from a diseased person and create embryos from them by resetting the clock of the cell nuclei. About one week into the process, stem cells, which are capable of differentiating into any type of cell at all, would be available for therapeutic use. Take the example of an individual suffering from Parkinson's disease. From cells of this patient an embryo could be created using nuclear transfer, the embryo stem cells might then be induced to differentiate into nerve cells required to treat the disease. Such cells would be injected into the diseased area of the patient's brain with the aim of eradicating the disease. Using nuclear transfer to make cells to order in the way just described is a likely though distant possibility. However, were it to become possible, in addition to Parkinson's it could be used in the treatment of diabetes, leukaemia, muscular dystrophy, and other diseases associated with damage to a single cell type against which humans have scant protection.

Although it is true that the process just sketched involves creating a human embryo, we must remember that at that stage in its development, the embryo would be comprised of 250 cells only and thus there is no question at all of human sentience. Despite the fact that this 250-cell embryo is a potential person, I personally would endorse its use in the treatment of otherwise untreatable disorders. The $64,000 question, of course, is whether we shall be able to discover a way to go from differentiated cells to undifferentiated cells without creating an embryo at all. The birth of Dolly and the other cloned animals shows that it is possible, by making an embryo, to take a differentiated cell and reverse all of the changes in the function of the nucleus which were required for the development of that specific type of cell. My view is that, in time, it will be possible to achieve this reversal without making an embryo.

Conclusion

Some of what I have envisaged above may ultimately prove impracticable. However, I do think that we have good reason to be optimistic about many of the possible applications of nuclear transfer technology discussed. It is entirely natural for us to want to explore nuclear transfer further. Although I believe that our research should be very ambitious and curious, I also believe that we should exercise extreme caution about how we use this new technique. To this end I think it utterly essential that wider society and not scientists such as myself, clinicians, patients, or drug companies, decide how nuclear transfer technology is to be developed and exploited. We, as a society, must ensure that this research is carried out within a set framework on which there is a broad consensus.

Cloning and Individuality

R. L. Gardner

Organisms that are genetically identical constitute a clone. The essence of cloning is to produce new individuals without the intervention of sexual reproduction, which generates diversity. Cloning is accomplished naturally by the repeated production of new individuals directly from parts of existing ones by such processes as cell division and budding. Particularly in microbes, plants, and lower animals, clones can be very large indeed. Cloning also occurs naturally in man as a result of the unexplained development of more than one fetus from a single fertilized egg. Whilst two genetically identical fetuses are the most common outcome, triplets, quads, and even quins can originate in this way. With increasing use of ultrasound imaging it is now clear that such monozygotic twins are more common during pregnancy than at birth. Thus, competition between intimately associated fetuses must sometimes lead to the demise of one, not infrequently at so early a stage that no trace of it remains when the survivor is born. Deliberate subdivision of early conceptuses would thus constitute a method of cloning that closely

I wish to thank Mrs Ann Yates, Professor Chris Graham, and Professor Peter Bryant for help in preparing this manuscript, Dr John Bassett for sharing his personal experience as an identical twin, and the Royal Society for support.

mimicked a natural process. However, given the greater hazards to mothers of carrying twins, such a practice would be of dubious value, especially since, unlike with cloning an adult or child, one would be reproducing a genetic constitution whose attributes were essentially unknown. This is why transplanting a nucleus from a cell of an existing individual to an unfertilized egg deprived of its own nucleus has been the primary focus of interest in the debate on human cloning. This approach has variously been termed nuclear transplantation, replacement, or substitution, and is the mode of cloning with which the rest of this commentary is concerned.

In his presentation, Dr Wilmut concentrated on the potential of cloning in mammals for enhancing the scope of biotechnology rather than the more emotive issue of its possible extension to man. It is the latter that prompted the global flurry of activity among politicians, legislators, and pressure groups immediately following the announcement of the birth of Dolly the sheep. The prospect or spectre of human cloning has, of course, been raised many times, but never before has it excited so much interest. What is different this time is that it was achieved in a species to which we are not too distantly related, and using an adult rather than an embryo as the donor. Furthermore, progress in assisted reproduction has removed many of the technical obstacles to applying this method of cloning to humans.

Numerous pronouncements have been made recently about whether cloning by nuclear transplantation should be permitted in certain circumstances or proscribed altogether. Before exploring an aspect of this issue which seems to have been largely neglected in recent debate, it is perhaps useful to consider why nuclear transplantation or substitution was done in the first place since, in this as in other endeavours, scientists are too often portrayed as people who simply tinker with Nature in an unthinking way.

The technique of nuclear transfer was introduced primarily to address basic biological issues and the concept of cloning, which so exercises people now, only emerged thereafter. Fertilization can be regarded as the most common form of nuclear transplantation that occurs naturally since the sperm is typically very poorly endowed with other cellular components relative to the much larger egg and, in many species, only its nucleus is necessary for subsequent development. Early this century, fertilization of nucleus-free fragments of eggs of one species with sperm from another was used to investigate the relative roles of the nucleus versus the rest of the cell in determining the pattern of the early embryo, particularly in sea urchins. Nuclei were first transplanted between cells experimentally in the humble amoeba, again with the aim of elucidating the role of cytoplasm and nucleus at a time when the chemical nature of genes was still uncertain.[1] Shortly thereafter, the technique was taken up in the frog to address a fundamental question concerning the process of embryonic development.[2] This was: 'What happens to the genes that are not required to function in cells when they become specialized?' e.g. genes such as those for pigmentation, or for insulin or antibody production in muscle, nerve, or kidney cells. Are they switched off irreversibly through chemical modification, rearrangement, or even deletion, or are they retained in a silent but potentially functional state?

That very sophisticated specialization of cells can occur without loss or irreversible modification of their genes is clear from the fact that eggs and sperm, which are arguably the most highly specialized cells in the body, can none the less produce a new individual after being brought together at fertilization. However, this could simply mean that germ cells, which are usually set aside early in development, are exceptional in retaining a capacity that the ordinary cells of the body, the somatic cells, have lost. The critical experiment was therefore to take nuclei from specialized somatic cells and inject them into unfertilized eggs whose own nuclei had been removed. If the donor nucleus could support completely normal development of a new individual, then it clearly had not suffered any loss or rearrangement of its genes during the course of specialization of the cell from which it was taken. Unfortunately, though clear-cut in principle, such experiments in frogs failed to provide a decisive result. While nuclei from embryonic cells could support the development of enucleated eggs into adults, those from obviously specialized cells gave, at best, only tadpoles. This prompted a lively and protracted debate, which is still not entirely settled, as to why this was so. One faction took this as evidence that genetic material is indeed altered during cellular specialization; another attributed it to damage in this material after transplantation because, when taken from a specialized cell, it had insufficient time to adjust to its new environment before being required to replicate itself.

One of the advantages of doing such experiments in mammals is that the pace of development is initially much slower than in frogs so that the likelihood of nuclei from specialized cells incurring damage following transplantation is lower. However, although attempts to do nuclear transplantation in mammals began nearly thirty years ago, it is only recently that they have met with any success. This is due in large measure to the greater technical problems posed by working with eggs that are much smaller than those of frogs. However, until Dolly, the only mammals that had been obtained were with nuclei derived from embryonic or fetal cells. The signal feature of Dolly is that the nuclear donor was an adult. Since relatively unspecialized stem cells persist in the adult, particularly in tissues like mammary epithelium which was the source of the transplanted nuclei, nothing definitive can be said about the status of the donor cell. The key consideration for most people is simply that Dolly provides evidence that cells from an adult can be used for cloning in mammals, thereby raising the prospect that this may now be feasible in man.

At present, it is important to differentiate between what is feasible and what is practicable because the reason why the process leading to the production of sheep like Dolly was so inefficient, even with fetal or embryonic rather than adult donor nuclei, remains unclear. Dolly emerged as the sole survivor to birth following successful transplantation of nuclei to 277 enucleated eggs.[3] Why was the success rate so low? One view is that with better understanding of the changes that donor nuclei have to undergo, and the consequent introduction of the appropriate technical refinements, the success rate will increase. However, at this juncture it is not possible to dismiss the alternative possibility that nuclei of only occasional donor cells retain competence to support normal development. Once cells have become specialized they only require a modest subset of all their genes to function. Adverse changes occurring in the remainder may be of no consequence for survival of the cells and thus not eliminated by selection. Especially once an individual has reached adulthood, the proportion of cells carrying such changes might be substantial. At present this remains largely an uncharted area. Regardless of the explanation for the present very low rate of success, it is important to ask how much more efficient it must be before it is held to be practicable in man. Here, one also has to consider the currently relatively low efficiency of *in vitro* culture and replacement procedures which would be necessary adjuncts to nuclear substitution.

Aside from the questions of what is feasible and what is practicable (see Alan Colman's comment on Putnam in Chapter 1) an important issue that has been overlooked by many in the debate over human cloning is what the outcome of cloning would be. Whatever the motivation, there seems to be a general expectation that the result would be one or more exact replicas of the nuclear donor. This expectation is clearly implicit in the suggestion that cloning might replace a cherished child or other close relative or partner. Obviously, there would be little point in going through the complexities of transplanting nuclei into enucleated eggs rather than conventional repro-duction unless this were so. The important question that does not really seem to have been addressed in the spirited, ongoing debate that was stimulated by Dolly is whether, on the basis of available biological knowledge, this expec-tation is warranted. In other words, if Beethoven or Einstein were available as nuclear donors how closely would their clonal descendants resemble them? That there would be a strong, if not striking, physical resemblance is hardly to be doubted in view of the generally close similarity between identical twins in this regard. However, the notion that cloning would enable perpetuation of the constellation of higher mental attributes which serve to distinguish human individuals is patently absurd. So why does this expectation endure? An important factor is the marked tendency, particularly in the popular literature, to emphasize the similarities rather than the differences between monozygotic twins, especially in cases where they have been reared apart.

Even physical resemblance is not always sufficiently strong to enable unequivocal classification of twins of like sex with separate placentae as mono- or di-zygotic without recourse to DNA testing. With regard to the question of the extent to which higher mental attributes are determined genetically we are confronted with the nature–nurture debate which seems to continue unabated in both passion and vitriol. Nevertheless, even the most ardent genetic determinist would not assert that any depend wholly on our genetic endowment.

Protagonists of the view that intelligence is a largely inherited attribute set great store by comparisons of identical twins reared apart versus unrelated individuals reared together.[4] However, others emphasize the importance of identical twins sharing a common uterine environment before birth regardless of whether they are together or apart thereafter. Some subscribe to the view that competition between identical twins *in utero* may tend to make them dissimilar. Such competition is likely to be more severe between later forming twins which share a common placenta than in earlier forming ones with separate placentae. However, recent evidence suggests that twins with separate placentae may be more dissimilar in IQ than those with a shared one.[5]

Thus studies in monozygotic twins clearly demonstrate that while people who share the same genetic constitution may be very similar in many respects, they nevertheless differ sufficiently to leave no doubt about their individual identity.[6] This point is illustrated particularly forcefully by the pair of conjoined twins, Eng and Chang Bunker, to whom the term 'siamese twins' was first applied. Born in Thailand (then Siam) in 1811, these twins remained attached to each other for the entire 63 years of their existence. Despite sharing the same genetic constitution and egg, and also the same uterine and postnatal environments, they exhibited very distinct personalities, tastes, and inclinations.[7] Members of a clone cannot resemble each other more closely than such identical twins and will, in general, be expected to do so rather less for at least two reasons. First, not all genes reside in the nucleus, some being sequestered in small bodies called mitochondria. Because eggs contain very many more mitochondria than sperm, these non-nuclear genes are inherited from the mother rather than from both parents. The contribution that mitochondrial genes make to differences between individuals is still far from clear, though recent studies have implicated mutations in such genes in a variety of diseases. Hence unless nuclear transplant hosts develop from eggs from the mother of the nuclear transplant donor, they may differ in their endowment of mitochondrial genes. Second, by virtue of their disparity in chronological age, nuclear transplant recipients and donors will invariably experience different environments both before and after birth, even if borne by the same individual. The importance of the uterine environment for susceptibility to disease later in life has only recently begun to emerge.[8] Whether the fact that human fetuses are sensitive to sound is of significance for their cognitive development has yet to be explored. Regardless, the prospect of accurately matching the postnatal environment of nuclear transplant recipients with that of donors would appear to be particularly remote. Even if this could be achieved, there is no reason to expect cloned humans to bear more than a haunting resemblance to the individual with whom they share genetic identity.

One can, therefore, legitimately question whether any compelling case could be made for producing adult humans via cloning rather than through sexual reproduction. Much easier to justify would be the use of nuclear transplantation into eggs to obtain specialized cells in culture which, by virtue of being genetically identical to the donor, could provide the latter with reparative grafts that would not suffer rejection. It would be regrettable if, in all of the controversy surrounding Dolly and her origins, the wider community lost sight of the benefits nuclear transfer techniques might confer on humankind.

Who Should have Access to Genetic Information?

Bartha Maria Knoppers

No discussion of a 'should we' question can ignore an even more preliminary one, a question that sets the stage for any present and future resolution of an issue, that is, 'where are we?' Obviously, this series, The Genetic Revolution and Human Rights, is meant to do just that. Nevertheless, permit me a few introductory remarks on the relationship between that fundamental exercise and the issue of access to genetic information.

As the various topics in this series illustrate, we are now in an era of bioengineering where, in fact, human genetics is just a small part. Reproductive technologies have thrown the 'begat-begat', generational model of the genetic lottery into disarray. The hierarchical, species model with human beings at the top followed by animals and plants has also been shaken as we discover new homologies between all species, use transgenic plants and animals as vectors for pharmaceutical products and vaccines, and bioengineer pigs to produce human organs. These radical transformations have launched us into a *biosociety*, or according to its critics—a *biotechnocracy*.

My argument below is not with the 'rightness' or 'wrongness' of this fact. Rather it is a plea for policy-making that does justice to its complexity. Mechanistic responses based on deterministic interoperations of this phenomenon rather than adaptive, dynamic, and flexible ones guided by normative principles, will ultimately harm the very human subjects and the society they are meant to serve. I would also further caution against any policy which does not situate these issues within the larger consideration of our role and responsibilities in this ecosystem. The errors and shortsightedness of our failure to respect the environment and other species are all too evident. Now that we have begun fashioning approaches for the new human genetics, humility, caution, and foresight based on knowledge are ethical prerequisites for intervention. Moreover, as globalization continues, possible responses to the issues need to emerge both nationally and internationally within a framework of shared principles.

Turning now to our subject, in this 'biosociety' or 'biotechnocracy', rapid advances in genetic research will ultimately result in affordable and more pervasive testing. Indeed, not only have we moved from tests for the rare, monogenic conditions to the discovery of genetic factors in common multifactorial diseases, but the development of 'DNA biochips' will allow testing for hundreds of conditions at a time. With the standardization of this technology, a single sample of DNA (found in every cell in the body) will provide information on the present and future health of a person and thus, necessarily, that of fellow family members. This latter characteristic of genetic information is worth mentioning. Genetic information not only has historical, eugenic connotations, and can be socially stigmatizing, it also involves one's parents, siblings, and children. When transported outside the highly confidential confines of the physician–patient relationship, it acquires roles and meanings that can affect the socio-economic survival and the relationships of the individual with his or her family.

The topic 'Who Should Have Access to Genetic Information?' has been hotly debated for over a decade. Yet, the first tentative solutions (if any) are only just beginning to appear in government and industry policy with regard to insurance and employment as well as in professional guidelines concerning communication of medical information to at-risk family members. Moreover, medical information itself is being increasingly 'geneticized'. The 'gene of the week' phenomenon together with the trend towards genetic reductionism—that is: gene = disease; disease = person; person = gene—will make it difficult to distinguish between genetic information and medical information. If so—and this may well be the proper characterization—are we pursuing or finding solutions in insurance and in employment when both the problem and the possible solutions lie elsewhere? Furthermore, does such information create new obligations between family members?

In 1991, Paul Billings, a geneticist critical of genetic discrimination, argued for a

[r]estatement that individuals have freedom of choice in personal health matters, the right to work and to conduct this pursuit in a safe place, the right to privacy, and the right to certain entitlements, including health care and the ability to insure the economic safety of the family. [This restatement] needs to accompany the 'genetic revolution'.[1]

Are insurance and employment a privilege, or a right, as just described? Does 'freedom of choice in personal health matters' include withholding genetic information from at-risk relatives?

To answer these questions, we must briefly describe the reasons for public disquiet before addressing the arguments both for and against the use of genetic information by insurers and employers. In the second part of this chapter I will discuss the possible emergence of a duty to warn at-risk relatives. The main thrust of my discussion, however, will examine the larger policy issues of solidarity, equity, and mutuality in modern societies, where 'congenital bad luck' can now be exposed and known even before it manifests itself.

1. Socio-Economic Harms?

Public concern with the ethical, legal, and social implications of access to genetic information by insurers and employers can be traced to three sources.

- First, employment and insurance are two of the most tangible ways in which genetic information may be used to the detriment of individuals. Employers and insurers who have genetic information about individuals are able to discriminate based on genetic factors, thereby denying individuals an opportunity to earn a livelihood and provide for the financial security of themselves and their families.

- Second, individuals' legitimate concerns about genetic-based discrimination frequently affect their health decision-making. Already, many individuals who are at risk of genetic disorders forgo genetic testing or participation in research because they fear the results will be obtained by their employer or insurer, thereby causing them either to lose or never to gain access to employment or insurance.

- Third, the disclosure of genetic information to employers and insurers raises important concerns about the privacy and confidentiality of genetic information, including the psychological and social consequences that flow from these disclosures.

Insurers and employers, however, will argue that it is 'business as usual' and that they should be privy to such information. Indeed, for years insurers have been using actuarial tables to assess the risks and compute the rates of individuals applying for life, disability, and additional health insurance. The advance of increased genetic testing in the clinical setting, however, creates the possibility that individuals who learn of their genetic risks would attempt to obtain substantial amounts of life insurance at standard rates without disclosing these risks. This behaviour is known as adverse selection or anti-selection, in which those at greatest risk seek the highest levels of coverage.

Thus the concerns of insurers are understandable. Insurance companies could not maintain business by selling large insurance policies to individuals who recently learned that they carried, for example, the gene for Huntington's disease or similar lethal, late-onset disorders. Unless the insurers have access to the same information as the applicant, they are at a disadvantage. They therefore argue that they must have the right not only to gain access to genetic information in medical records, but also to perform their own genetic tests where they deem this appropriate.

Employers also have an economic interest in the health of their employees. When employees are in poor health they are less productive, have a higher rate of turnover, are more likely to use sick leave, and are more likely to suffer injury and illness on the job. And their illness may adversely affect the morale of co-workers and customers. Genetic information could be used to predict which asymptomatic individuals are likely to develop late-onset monogenetic disorders (e.g. myotonic dystrophy) as well as those who are at increased risk of multifactorial disorders (e.g. cancer and cardio-vascular diseases). In some unusual situations, genetic factors (e.g. α-1 antitrypsin deficiency) may predispose individuals to occupational disease (respiratory disorders) when combined with occupational exposures (dusty conditions). In other rare situations, a genetic disorder (e.g. Marfan syndrome) may suddenly incapacitate an employee, causing risk of serious injury to the employee, to co-workers, or to the public.

An employer's economic incentive to exclude from the workplace individuals who are more likely to become ill is largely the same for genetic disorders as it is for non-genetic disorders. In fact, employers have the right to select the most productive applicants within the twin constraints of human rights and unfair labour practices legislation. The difference with genetic disorders is that individuals can be identified before the onset of symptoms. With only a few notable exceptions (e.g. HIV infection), this is not possible for non-genetic disorders.

Ironically, access to genetic information and testing may also be in the interest of the applicant for insurance or employment. Such information will enable insurers to 'establish more detailed individual risk assessments than ever before. Insurance pools will be reduced and more detailed underwriting through genetic tests will become cheaper and thereby more financially interesting'[2] for both the insurer and the 'genetically healthy'. Similar arguments could be made by the 'genetically healthy' for chances at employment and promotion.

There looms the possibility of systematic discrimination against the 'asymptomatically ill'[3] and the possible creation of a new special underclass. As social security and universal health care unravels, and as access to insurance becomes a necessary economic good for the obtaining of other goods, such as a home or a car, and as employment pensions begin to replace reliance on government schemes, what legal protection is available?

Legal protection is a double-edged sword.[4] (We have seen that both employers and insurers can 'legitimately' discriminate and select using *bona fide* job requirements and actuarial data respectively.) Currently, human rights anti-discrimination legislation does not include 'genetic conditions' in its list of prohibitions. Some would argue that it could be included under the broad category of handicap or disability. Or could this lacuna be somewhat remedied by the addition of 'genetic conditions' to the list? But the new genetics exposes asymptomatic individuals, not those with manifest conditions who would already be protected, so this addition of genetic conditions would not be helpful. Perhaps the inclusion of the phrase, 'believed to have' would cover the perception of handicap, that is, where the person is treated as already ill since it is this perception of handicap that causes unfair discrimination. Or, finally, genetic-specific legislation could be adopted forbidding the use of genetic information in certain sectors such as insurance and employment.

There is, however, a danger in singling out genetic information and conditions. Indeed, such a distinction may ultimately be more stigmatizing than the original condition. Amendments to human rights legislation with the addition of 'genetic condition', or the adoption of specific statutes, then, are likely to be problematic in that they could further contribute to public perceptions of various genetic traits as 'abnormalities'. It is this very perception of abnormality, which has also contributed to the hesitancy of family members to share information amongst themselves, that is our second area of concern.

2. Familial Harms?

At a time in history when the genealogical has been slowly replaced by the consensual, new family forms, the advent of genetic testing is forcing its recreation. When direct testing is not possible, family members are needed for pedigree and linkage studies in order to estimate individual risk. Thus, irrespective of these new sociological configurations, genetic research with its familial recruitment may yield genetic information that is important to other blood relatives. The fact of being tested or of

participation in research or not; the decision to refuse to warn at-risk family members or to withdraw [from research] or the failure to provide for access [to DNA samples] after death, all affect the interests of present and future family members. These shared biological risks create special interests and moral obligations . . . that may outweigh individual wishes.[5]

Should genetic information be considered familial property? Do personal autonomy and the medical duty of secrecy trump a possible duty to warn?

While there may be a moral obligation, currently there is neither a legal duty to warn between blood relatives nor on the part of physicians. The only exceptions in the case of the latter arise in situations of child abuse, gun-shot wounds, and so on. While some would agree that the creation of such a duty to warn could lead to fear of testing or participation in genetic research, others would argue for the recognition of a duty to rescue where treatment or prevention is available. Like insurance and employment then, the question of intra-familial communication raises the issue of risk-sharing. Who carries the burden, under what conditions, and based on what values?

3. Value-Creation or Reaffirmation?

It is important to remember that insurance is based on

the complementary principles of solidarity and equity. Solidarity means that the population as a whole or in broad groups, shares the responsibility and the benefits in terms of costs, while equity in the context of insurance means that each individual's contribution should be roughly in line with his or her *known* level of risk.[6]

Furthermore, both employment and insurance have traditionally been seen as commodities in the marketplace. Other than the marketplace criterion of fairness, that is, as we have mentioned, the filter of 'legitimate' discrimination, both are essentially marketplace goods. However, 'to the extent that our society decides that all insurance companies [and employers] should have a redistributive role, the nature of the goods changes and the rules of equity of the *marketplace* might no longer apply'.[7] Likewise, to the extent that our society decides that there exists at least a minimal obligation of mutuality between at-risk family members, individual autonomy and medical secrecy can be tempered. Solidarity in risk-spreading and mutuality in risk-sharing are, then, values to be reaffirmed and promoted but not without further qualification.

First, the future adoption by insurers and employers of additional obligations as corporate citizens should not be at the expense of a further dismemberment and devolution of the responsibility of the state. Social security and universal health care must not be further undermined. Secondly, any exceptions to the sacrosanct principles of personal autonomy and medical secrecy must be clearly delineated and strictly curtailed. Thirdly, the approach taken must be one that reflects the complexity of the issues. Thus it must be dynamic and flexible. This rules out the adoption of incremental genetic-specific legislation and argues in favour of multi-dimensional, evolutionary, and adaptive approaches. This means including both state regulation and professional self-regulation.

Within these guiding principles and procedures, what reforms can we envisage? Taking the issue of access by insurers and employers and family members in turn, it will become evident that both the 'normalization' of genetic information and the notion of our citizenry are at the core.

Conclusion: Towards an Integrated Approach

Crucial to any reforms at the national level is the adoption and integration of the guiding principles of both UNESCO's Declaration on the Human Genome and Human Rights and the Council of Europe's Convention on Biomedicine.

The Universal Declaration (adopted in November 1997) stipulates in Article 6:

No one shall be subjected to discrimination based on genetic characteristics that is intended to infringe or has the effect of infringing human rights, fundamental freedoms and human dignity.

And in Article 7:

Genetic data associated with an identifiable person and stored or processed for the purposes of research or any other purpose must be held confidential under the conditions set by law.

Similarly, the European Convention states in Article 11:

Any form of discrimination against a person on grounds of his or her genetic heritage is prohibited.

And in Article 10:

Everyone has the right to respect for private life in relation to information about his or her health.

Everyone is entitled to know any information collected about his or her health. However, the wishes of individuals not to be informed shall be observed.

In exceptional cases, restrictions may be placed by law on the exercise of the rights contained in paragraph 2 in the interests of the patient.

Finally, and importantly, Article 12 maintains:

Tests which are predictive of genetic diseases or which serve to identify the subject as a carrier of a gene responsible for a disease or to detect a genetic predisposition or susceptibility to a disease may be performed only for health purposes or for scientific research linked to health purposes, and subject to appropriate genetic counseling.

Keeping these principles in mind, it is obvious that fundamental to all three areas—insurance, employment, and a possible duty to warn—is the classification of genetic information, not as distinct or different, but as sensitive, medical information. If treated as such, the central consideration, then, is its use for the purposes of the health and well-being of the person and his/her family, through protection, promotion, and prevention. To that end, there is an absolutely urgent need to reinforce and strengthen existing legislation on the confidentiality of medical and research data generally.

If we assume that medical and research data will be better protected, what are the implications for insurance? I propose that the following be adopted:

1. A continued moratorium on any request to an applicant to take a genetic test as a condition of offering him or her insurance;
2. During this moratorium, the insurance industry should develop accurate, scientifically validated predictive values for the future of genetic information in underwriting;
3. Further epidemiological studies should be conducted

to ensure actuarial fairness in such underwriting practices;

4. No access should be permitted to research records or to research results;

5. During this moratorium, the insurance industry should also consider creating risk pools with community rating without requiring access to individual information and base its calculations simply on currently obtained family histories. As an alternative, or in addition to the above, a minimal standard 'no questions asked' policy should be made available (this is already the case in the United Kingdom) or one proportionate to the social and financial circumstances of applicant (the approach taken in the Netherlands)—the preferable approach. For applications over that minimum or proportionate standard, access to medical records should be permitted;

6. Such access, however, should only be to certain categories of information and not to the whole medical record;

7. These general categories of access should be determined by a body of insurers, health professionals, and patients' associations working together;

8. National associations of insurers, health professionals, and patients' associations should also develop and adopt codes of practice and make them publicly available;

9. Finally, the moratorium should be lifted when the industry can demonstrate, to an outside expert body, the scientific accuracy and validity of its incorporation of predictive genetic information into its tables.

Turning to employment, assuming again that medical

and research data will be more adequately protected, the implications with respect to employment are as follows:

1. Anti-discrimination legislation should be amended to add to the enumerated conditions including handicap or disability, the words 'believes the person has or will have' so as to protect against discrimination resulting from perception of handicap or disability;
2. Pre-employment genetic testing should be undertaken only when it is scientifically proven by the employer to be job-related;
3. Specific and informed consent should be obtained prior to monitoring an employee and the nature and purpose of all procedures should be explained;
4. A national agency should designate which genetic tests are scientifically valid and accurate in a given employment context for safety and health purposes;

Finally, as concerns duties between relatives, reaffirming the professional duty of confidentiality and of personal autonomy and privacy but recognizing, nevertheless, the potential for harm to at-risk relatives, the implications for health-care professionals where the individual refuses to warn are that:

Special consideration should be made for access by immediate relatives. Where there is a high risk of having or transmitting a serious disorder and prevention or treatment is available, immediate relatives should be able to learn . . . their own status. These exceptional circumstances should be made generally known in both the institutional [professional] and the research relationship.[8]

Such an approach involves professional self-regulation,

moratoria, and amendments to human rights legislation all at the same time. Failure to adopt this multi-faceted approach could have certain consequences such as the adoption (as is already the case) of hastily drafted legislation that is either overly broad or too narrow and genetic-specific. This could irrevocably harm both the industry and citizens. Paradoxically, citizens who suffer from (as yet) non-genetic diseases may well claim discrimination if legislation is over-protectionist of genetic conditions. Industry itself could stall reforms if unjustifiably accused or constrained. In short, while a policy of zero-risk toler-ance runs contrary to an appreciation of the complexity of the human condition and life in society, this does not mean that those vulnerable to genetic risk must bear the burden of its miscomprehension alone in the socio-eco-nomic contexts of insurance and employment. Nor does it mean, however, that they can knowingly and intentionally withhold such information when it can be shown to prevent certain harm to an identified family member. If we expect and require industries to behave in a socially responsible fashion as corporate citizens so as to forestall for the present the untoward consequences of genetic uncertainty, surely the individual citizen should be held to the same standard!

Corporate solidarity and familial mutuality may then be but the first steps in a new form of justice whose principles of fairness are based on equity and knowledge. Ultimately, both industry and the individual must confront this challenge of modern citizenry. This modern form of citizenry evokes the ancient Biblical question: 'Am I my brother's keeper?' Surely the answer is 'yes' for both.

Bad Genetic Luck and Health Insurance

Justine Burley

Professor Knoppers locates three sources of public concern over access to genetic information: the potential for discriminatory uses of such information by insurers and employers; the fear that genetic-based discrimination will affect individuals' decision-making about health; and the implications of disclosure of genetic information for privacy and confidentiality in various domains including that of the family. Professor Knoppers covers much difficult ground and offers, in the process, many insights into the problems we are facing as well as sensible solutions to them. My own discussion will be limited to an exploration of a view suggested by Knoppers's proposals for how the use of genetic information in the insurance market should be constrained. It seems to me obvious enough that the potential misuse of genetic information by insurers is troublesome first and foremost because it is unfair that some people are worse off than others due to bad genetic luck. The question to which I seek to provide a partial answer in what follows is: 'How should government

I am grateful to Matthew Clayton, Sara Connolly, and Charles Erin for their helpful comments on earlier drafts of this commentary.

respond to inequalities between individuals' respective genetic endowments?'

The steady increase in the number and sophistication of genetic tests affords the possibility of detailed diagnostic and predictive genetic information about particular individuals. Professor Knoppers endorses (rightly) an approach to policy-making that is responsive to the complexity of this phenomenon, one that is integrated and which is guided by normative principles. The reforms to insurance practices that Knoppers proposes along with her affirmation of existing legislation are designed to ensure (1) that no one is denied insurance and (2) that the genetically 'unhealthy' are not, relative to people who have non-genetic disabilities, disadvantaged in the insurance market.[1] In the case of health insurance this latter aim is a minimal consistency requirement which I think can be improved on.

My central contention in what follows is that society at large should share the costs of the bad genetic luck that its individual members suffer[2] by assigning to the state the redistributive role that Knoppers envisages being performed (to some extent) by insurance companies. Even were Knoppers's reforms adopted it would still be the case that a mandatory state health insurance scheme could secure more fairly coverage for victims of bad genetic luck.[3] Ideally, society should be a sort of 'People's Republic of Underwriters' (PRU). This is not to say that I deem private insurance to have no place at all, only that I believe it just for the state to enforce redistribution from the genetically lucky to the genetically 'under endowed'. In support of this view I shall defend three claims. First, equity is best achieved by requiring that everyone, including the genetically 'healthy', pay out money in insurance, and that the rates paid are not indexed to a person's known level of genetic risk.[4] Second, differentially priced premiums are also unfair because whether or not many genetically based diseases actually develop depends on the influence of external environmental factors, the creation of and exposure to which is often not controllable by particular individuals. Finally, a mandatory state scheme would, like Knoppers's own proposals, remove one powerful disincentive people might have to undergoing genetic testing: fear of discrimination by insurers. To give shape to these claims I shall now describe a hypothetical society in which insurance practices are governed by Knoppers's reforms.

In this hypothetical society a battery of tests from birth will reveal, to a high degree of accuracy, an individual's risk of developing any and all genetic conditions. For simplicity's sake let us assume that the only cause of illness is genetic in origin, and that all members of the population are risk averse to exactly the same degree. Let us also assume that the citizens here fall in equal numbers into one of three distinct genetic categories: 25 per cent of the population test negative for all (except ageing) genetic-specific conditions ('negatives'); 5 per cent test positive for a seriously debilitating mono-genetic disorder ('monos'), the onset of which is sure; and the remaining 70 per cent test positive for various poly-genetic predispositions and susceptibilities ('polys') which will only become manifest in those individuals who are exposed to certain known health-endangering environments (e.g. urban).

Other things being equal, what relative amounts of

coverage would be sought by members of each cohort? The 'negatives' would not want any at all, the 'monos' would seek extensive coverage, and 'polys' would either want no, some or extensive coverage—the amount desired by each 'poly' will depend on the frequency of his/her exposure to certain environments.

It is to the benefit of insurance companies to incorporate scientifically accurate genetic predictive data into their actuarial tables in order to distinguish between risk groups. Knoppers would permit insurers in this society to solicit from insurees information about their known level of genetic risk though she stipulates that risk profiles may not be a condition of purchase for all available premiums: insurers must offer a variety of differentially priced packages including minimal coverage 'no questions asked' policies and no one may be denied access to insurance.[5] Thus Knoppers's vision is of an insurance industry which adheres more closely to ideals of solidarity and equity than is currently the case.

But even if it is redesigned in this way, private insurance cannot secure the goal of equity adequately: 'low risk' individuals may not insure while those at risk may be charged higher rates than others for the same coverage or might only be able to insure for minimal coverage. Our hypothetical society clearly illustrates the problem connected to non-insurance. The 'negatives' (25 per cent of the population), for example, purchase no insurance because available genetic tests and family history give a sunny picture of their health status. However, in failing to share at all in the costs of bad genetic luck experienced by others, they act unfairly. This is because, other things being equal, the 'negatives'' good fortune is simply due

57

to their having had good luck in the genetic lottery. The equity problem is also exacerbated by differentially priced policies. For example, insurance companies may offer 'polys' who unavoidably shift between innocuous and health-endangering environments, coverage at a lower rate than that charged to 'monos' and also to those 'polys' constantly exposed to health-compromising environments. As with non-insurance, differential pricing has the effect of disadvantaging further those who are already worse off due to bad genetic luck.

An additional problem illustrated by our hypothetical society is that posed by external environmental factors. For a different reason than the 'negatives', a proportion of the 'polys' purchase no insurance. These individuals do possess offending genes but because they will not, unlike their 'poly' counterparts, be exposed to condition-aggravating external environments they are not at risk of developing disease. The existence of health-endangering environments is, at least in part, the responsibility of society as a whole. And, the choices of where to work (think of the unemployment rate) and live (think of children) cannot in many cases reasonably be thought 'free' ones—such choices are much constrained by the vagaries of the marketplace and social circumstance. Therefore it is unfair that some individuals pay more in insurance for the same coverage afforded others or only have access to minimal coverage, when they cannot avoid exposure to these environments.

The unfairnesses just identified are, of course, characteristic of the insurance market in general, but it does not follow from this that we should blithely accept this situation, nor that a more just response is unavailable. It is morally incumbent on us as a community to mitigate the bad genetic luck which victimizes only some of us. Good health is an essential ingredient of all individuals', otherwise very different, respective conceptions of what it means for a life to go well. Were we to make health insurance mandatory we could address the problem of bad genetic luck fairly: we would guarantee equity by denying the option of no or little insurance to individuals whose known level of genetic risk is negligible and we would secure adequate coverage for those who become ill.[6]

To these arguments we can add a third reason in support of mandatory insurance: it would remove one of the main worries that people might have about undergoing genetic testing, namely, the possibility of discrimination by insurers. As noted by Knoppers, reluctance to be tested can not only impact adversely on one's own health but also on that of family members. If I am not tested for fear of discrimination by insurers, I lack detailed knowledge of my genetic predispositions and susceptibilities. This information might be valuable for decision-making about medical treatment and/or my lifestyle. It might also be medically useful to people genetically related to me (and, if they are planning a family, to their spouses). Knoppers's reforms also avoid this problem but, as I have argued, they do not appeal in other ways.

Professor Knoppers does stress that her proposals must not be accompanied by any further erosion of public health-care provision. It is likely, however, that such an erosion would be more apt to occur were an increased redistributive role for private health insurers promoted. Thus we should concentrate our efforts on bolstering the existing state apparatus by implementing insurance arrangements targeted specifically at mitigating bad luck with respect to health (genetic or otherwise).

Clones, Genes, and Human Rights

John Harris

The birth of Dolly, the world famous cloned sheep, has had an extraordinary impact on many dimensions of our lives, both intellectual and real. It has fuelled debate in a number of fora: genetic, scientific, political, moral, journalistic, and literary. It has also given rise to a number of myths, not least among which is the myth that Dolly presents a danger to humanity, the human gene-pool, genetic diversity, the ecosystem, the world as we know it, and the survival of the human species.

Dolly, or the technology by which she was created, raises a number of questions about human rights and how we are to understand the idea of respect for these rights and for human dignity. It is these questions that are the subject of this lecture. Cloning of the sort used to create Dolly has raised three main sets of questions which are vitally relevant to all who are interested in human rights and human freedoms and I shall try to say something about each of these.

I must thank Justine Burley for reading, commenting on, and revising several drafts of this chapter and some of the others from which parts of it are drawn. I should also like to thank Christopher Graham and Pedro Lowenstein for many constructive comments.

First, the birth of Dolly provoked a hysterical reaction, the burden of which was that she represented an attack on human dignity and on human rights and values of unprecedented urgency and severity. Secondly, the legislative and regulatory response to this hysteria itself raised huge questions of vital significance to all concerned with upholding human rights. Finally, whatever one might *feel* about Dolly and the possibility of human equivalents, our freedom to act on these feelings may, as I shall argue, be circumscribed by other values that we hold, such as the conception of human rights to which most democratic societies are committed and which is presupposed by democracy itself. Thus, far from outlawing Dolly and her kind, it may actually be required of us, if not to welcome a human Dolly, at least to afford her, and her would-be creators, tolerance and respect. This tripartite family of concerns generates three specific questions which I shall try to answer in what follows. They are:

1. *Does the creation of Dolly constitute an attack on human rights and dignity?*
2. *Does the legislative response to Dolly constitute an attack on human rights and dignity?*
3. *Are human Dollys protected by existing conventions on, and assumptions concerning, human rights?*

1. Does the Creation of Dolly Constitute an Attack on Human Rights and Dignity?

When Dolly's birth was reported in *Nature* on 27 February 1997,[1] the hysteria to which I have referred was immediately unleashed. The President of the United States called immediately for an investigation into the ethics of such procedures[2] and announced a moratorium on public spending on human cloning. Clinton's investigation has now reported and, commenting recently, President Clinton said 'there is virtually unanimous consensus in the scientific and medical communities that attempting to use these cloning techniques to actually clone a human being is untested and unsafe and morally unacceptable'.[3] Recently, Members of the European Parliament (MEPs) demanded that each EU member state 'enact binding legislation prohibiting all research on human cloning and providing criminal sanctions for any breach'.[4]

Perhaps in order to forestall or even 'manage' such reactions, the group which produced Dolly had apparently consulted a public relations firm to guide them on how to present what they realized would be a sensational achievement. In the light of events, they should possibly also have consulted a bioethicist or a bioethics centre.

Even commentators from whom a more considered approach might have been expected were panicked into instant reaction. Dr Hiroshi Nakajima, Director General of the World Health Organization, said: 'WHO considers the use of cloning for the replication of human individuals to be ethically unacceptable as it would violate some of the basic principles which govern medically assisted procreation. These include respect for the dignity of the human being and protection of the security of human genetic material.'[5] WHO followed up the line taken by Nakajima with a resolution of the Fiftieth World Health Assembly which saw fit to affirm 'that the use of cloning for the replication of human individuals is ethically unacceptable and contrary to human integrity and morality'.[6]

Federico Mayor of UNESCO, equally quick off the mark, commented that 'Human beings must not be cloned under any circumstances'.[7] Finally, on 3 December 1997, UNESCO published a so-called *Universal Declaration on the Human Genome and Human Rights*, Article 11 of which announced: 'Practices which are contrary to human dignity, such as reproductive cloning of human beings, shall not be permitted.'[8] In a staggeringly complacent preface Federico Mayor states: 'The uncontested merit of this text resides in the balance it strikes between safeguarding respect for human rights and fundamental freedoms and the need to ensure freedom of research.' However, Mayor is more than a little misleading when he reports in the Preface to the *Universal Declaration* that it 'was adopted unanimously and by acclamation'. We know[9] that the statement defining cloning as 'contrary to human dignity' was a late addition by UNESCO to the text originally produced by UNESCO's International Bioethics Committee. Moreover, one of the members of that committee, the distinguished molecular geneticist Michel Revel, who was also Israel's representative to the General Conference of UNESCO, has reported that 'several delegations proposed not to rush in condemning any particular technique, including cloning'.[10]

More recently, Mayor has again expressed his objections to human cloning.[11] He believes opposition to a ban on reproductive cloning 'is based on two main types of argument. One defends individual "rights" to clone, the second scientific freedom'. I should state at once that my opposition to the ban is based on neither of these. It is, to be sure, based on freedom, but not on scientific freedom. I believe, and I shall return to this argument in the final section of this paper, that human liberty may not be abridged without good cause being shown. It is the argument of this paper that no one, not even UNESCO, has shown anything approaching good cause.

The European Parliament also rushed through a resolution on cloning (Paragraph B of) the preamble of which asserted,

[T]he cloning of human beings . . . cannot under any circumstances be justified or tolerated by any society, because it is a serious violation of fundamental human rights and is contrary to the principle of equality of human beings as it permits a eugenic and racist selection of the human race, it offends against human dignity and it requires experimentation on humans.

And which went on to claim (in Clause 1) that 'each individual has a right to his or her own genetic identity and that human cloning is, and must continue to be, prohibited'.[12]

These statements are almost entirely devoid of argument and rationale. There are vague references to 'human rights' or 'basic principles' with little or no attempt to explain what these principles are, or to indicate how they might apply to cloning. The WHO statement, for example, refers to the basic principles which govern human reproduction and singles out 'respect for the dignity of the human being' and 'protection of the security of genetic material'. How is the security of genetic material compromised? Is it less secure when inserted with precision by scientists, or when spread around with the characteristic negligence of the average human male? Those of mischievous disposition might be tempted to ask whether the sin of Onan was not perhaps to compromise the security of his genetic material?

Human Dignity

The idea and ideal of human dignity have been much invoked in these debates. Typical of appeals to human dignity was that contained in the World Health Organization statement on cloning issued on 11 March 1997: 'WHO considers the use of cloning for the replication of human individuals to be ethically unacceptable as it would violate some of the basic principles which govern medically assisted procreation. These include respect for the dignity of the human being.' Appeals to human dignity are, of course, universally attractive; they are the political equivalents of motherhood and apple pie. Like motherhood, if not apple pie, they are also comprehensively vague. A first question to ask when the idea of human dignity is invoked is: whose dignity is attacked and how? If it is the duplication of a large part of the human genome that is supposed to constitute the attack on human dignity, or where the issue of 'genetic identity' is invoked, we might legitimately ask whether and how the dignity of a natural twin is threatened by the existence of her sister and what follows as to the permissibility of natural monozygotic twinning? However, the notion of human dignity is often linked to Kantian ethics and it is this link I wish to examine more closely here.

A typical example, and one that attempts to provide some basis for objections to cloning based on human dignity, is that provided by Axel Kahn (a distinguished molecular biologist who helped draft the French National

Ethics Committee's report on cloning). Kahn invokes this principle in his commentary in *Nature* where he states:

The creation of human clones solely for spare cell lines would, from a philosophical point of view, be in obvious contradiction to the principle expressed by Emmanuel Kant: that of human dignity. This principle demands that an individual—and I would extend this to read human life—should never be thought of as a means, but always also as an end.[13]

The Kantian principle, invoked without any qualification or gloss, is seldom helpful in medical or bio-science contexts.[14] As I have argued in response to Kahn elsewhere, his formulation of it would surely outlaw, amongst other things, blood transfusions. The beneficiary of blood donation neither knows nor usually cares about the anonymous donor(s) and uses the blood and its donor(s) exclusively as a means to her own ends; the donor figures in the life of the recipient of blood exclusively as a means. The blood in the bottle has, after all, less identity, and is less connected with the individual from which it emanated, than chicken 'nuggets' on the supermarket shelf. An abortion performed exclusively to save the life of the mother would also, presumably, be outlawed by Kahn's understanding of Kant's principle.

Instrumentalization

This idea of using individuals as a means to the purposes of others is, particularly in the European context, sometimes termed 'instrumentalization'. The 'Opinion of the group of advisers on the ethical implications of biotechnology to the European Commission',[15] for example, in its statement on 'Ethical aspects of cloning techniques' uses this idea repeatedly. For example, referring to reproductive human cloning (in paragraph 2.6) it states: 'Considerations of Instrumentalization and eugenics render any such acts ethically unacceptable.'

Making sense of the idea of 'instrumentalization' is not easy! If someone wants to have children in order to continue their genetic line do they act instrumentally? Where, as is standard practice in IVF, spare embryos are created, are these embryos created instrumentally? Kahn has considered these objections but he has not done so adequately.[16] He reminds us, rightly, that Kant's famous principle states: 'respect for human dignity requires that an individual is *never* used . . . *exclusively* as a means' and suggests that I have ignored the crucial use of the term 'exclusively'. I did not of course, and I'm happy with Kahn's reformulation of the principle. It is not that Kant's principle does not have powerful intuitive force, but that it is so vague and so open to selective interpretation and its scope for application is consequently so limited, that its utility as one of the 'fundamental principles of modern bioethical thought', as Kahn describes it, is virtually nil.

Kahn himself rightly points out that debates concerning the moral status of the human embryo are debates about whether embryos fall within the scope of Kant's or indeed any other moral principles concerning persons; so the principle itself is not illuminating in this context. Applied to the creation of individuals who are, or will become, autonomous, it has limited application. True the Kantian principle rules out slavery, but so do a range of other principles based on autonomy and rights. If you are interested in the ethics of creating people then, so long as existence is in the created individual's own best interests, and the individual will have the capacity for autonomy, then the motives for which the individual was created are either morally irrelevant or subordinate to other moral considerations. So that even where, for example, a child is engendered exclusively to provide 'a son and heir' (as so often occurs in many cultures) it is unclear how or whether Kant's principle applies. Either other motives are also attributed to the parent to square parental purposes with Kant, or the child's eventual autonomy, and its clear and substantial interest in or benefit from existence, take precedence over the comparatively trivial issue of parental motives. Either way the 'fundamental principle of modern bioethical thought' is unhelpful.

In Chapter 1, the distinguished American philosopher Hilary Putnam reiterates the Kantian imperative employed by Axel Kahn. Putnam imagines a scenario in which cloning is widely used by ordinary people so that they can have children 'just like so-and-so'. Putnam claims that 'what horrifies us about this scenario is that, in it, one's children are viewed simply as objects, as if they were commodities like a television set or a new carpet. Here what I referred to as the Kantian maxim against treating another person only as a means is *clearly* violated' (p. 8).

Criticizing Richard Lewontin,[17] Putnam suggests that Lewontin, and by implication the arguments defended in this lecture, are confused over the meaning of the Kantian principle he has invoked. Lewontin had pointed out, surely correctly, that almost all commercial relations people have with one another are basically instrumental. Putnam attempts to remedy the alleged confusion with this illus-tration: 'Even when someone is one's employee, there is a difference between treating that someone as a mere thing and recognizing his or her humanity. That is why there are criteria of civilized behaviour with respect to employees but not with respect to screwdrivers' (p. 8).

Putnam is of course right to say that we do indeed have 'criteria of civilized behaviour' with respect to children. This is what distinguishes not only our (humankind's) attitude to children generally, but each parent's attitude to his or her own child in particular, from 'criteria of civilized behaviour' towards screwdrivers. There is no evidence for, and indeed no plausibility in, the supposition that if people choose to use a cloned genome in order to create *their own children*, that these children will not be loved for themselves, let alone not treated in a civilized way. We have noted that many people have children for a purpose; to continue their genes, to provide a son and heir, to create 'a sister for Bill', to provide for support in old age, 'because I've always wanted a child to look after', because our tribe or our ethnic group is threatened with extinction, etc. When, if ever, is it plausible to say that they are having children *exclusively* for such purposes?

Kant's maxim provides a plausible account of what's wrong with slavery, for example, and, if another were needed, it provides one of many objections to Nazi practices. Where it conspicuously fails to be of any help is when people use, or even think of, others *partially* in instrumental terms, as happens in employment, family relations, sexual relations, and almost any human context. As I have argued, it is almost never plausible to think that people, whose motives and intentions are almost always mixed and complex, could definitively be said to be treating others 'exclusively' as a means, unless, as with the Nazis, they treated them as slaves or literally as things.

It is therefore strange that Kahn, Putnam, and others invoke the Kantian principle with such dramatic assurance, or how anyone could think that it applies to the ethics of human cloning. It comes down to this: either the ethics of human cloning turn on the creation or use of human embryos, in which case as Kahn himself says 'in reality the debate is about the status of the human embryo' and Kant's principle must wait upon the outcome of that debate. Or it is about the ethics of producing clones who will become autonomous human persons. In this latter case, the ethics of their creation are, from a Kantian perspective, not dissimilar to other forms of assisted reproduction, or, as I have suggested, to the ethics of the conduct of parents concerned *exclusively* with producing an heir, or preserving their genes or, as is sometimes alleged, making themselves eligible for public housing. Debates about whether these are *exclusive* intentions can never be neatly resolved and are, in my view, ultimately sterile and unhelpful.

Putnam supplements his use of the Kantian principle with an interesting idea, that of a 'moral image' (p. 4). Putnam accepts that such images are plural and diverse and that 'people with different moral images may lead equally good moral lives' (p. 4). For Putnam any moral image must incorporate the Kantian principle which he believes is itself inspired by a moral image of autonomy 'our capacity to think for ourselves in moral matters' (p. 9). He then recommends an image of the family in which 'the good parent . . . looks forward to having children who will live independently of the parents not just in a

71

physical or an economic sense, but in the sense of thinking for themselves' (p. 9). This moral image has also to incorporate a 'willingness to accept diversity'. He then asks and answers a very important question:

But why should we value diversity in this way? One important reason, I believe, is precisely that our moral image of a good family strongly conditions our moral image of a good society. Consider the Nazi posters showing 'good' Nazi families. Every single individual, adult or child, male or female, is blond; no one is too fat or too thin, all the males are muscular, etc.! The refusal to tolerate ethnic diversity in the society is reflected in the image of the family as utterly homogeneous in these ways (p. 10).

Putnam continues:

Our moral image of the family should reflect our tolerant and pluralistic values, not our narcissistic and xenophobic ones. And that means that we should welcome rather than deplore the fact that our children are not us and not designed by us, but radically Other.

Am I suggesting then, that moral images of the family which depict the members of the ideal family as all alike, either physically or spiritually, may lead to the abominations that the eugenics movement contributed to? The answer is, 'very easily' (p. 12).

And concludes 'But perhaps one novel human right is suggested by the present discussion: the "right" of each new-born child to be a complete surprise to its parents!' (p. 13)

I am in great sympathy with much of this, indeed I have argued along the same lines myself.[18] For example, in discussing the practice of race matching in adoption I commented:

Why do so many people firmly believe that children should be like their parents, particularly in terms of their general colour and racial characteristics? It is difficult not to view this desire, and attempts to implement it, as a form of 'ethnic cleansing'. It smacks very much of the pressure that so many societies and cultures have put upon their members not to 'marry out' or, to put it more bluntly, not to mate with somebody of another tribe or race.

It is perhaps timely to press the question: why do we assume that the desire for a different-race child is racially motivated in some discreditable way, whereas the desire for a same-race child is not? If we are going to suspect people's motives, the desire for a child of the same race is surely as likely to be discreditable. It is, after all, societies which exclude different races that are assumed to be racist, not societies which welcome and celebrate diversity. Why should this not be true of families?[19]

The problem I have is with Putnam's interpretation of what follows from the attractive image that he presents. First, it does not follow from the fact that something is inconsistent with a moral image, that it should be banned, or controlled, nor does it follow that those who would be deviants may be punished.[20] Not only is Putnam right to say that there are many equally good moral images, there are also many equally good interpretations of the same moral image. I accept Putnam's image but, unlike Putnam, I conclude that parents should therefore have free choice in designing their children. As I have argued at length elsewhere in the context of parents using reproductive technology, to choose children's phenotypic traits:

Some people have objected that to choose the skin colour or racial features of children (insofar as these can be chosen—which is not very far) is an illicit form of parental preference. The phrase 'designer children' is often used pejoratively to describe the children of parents who are more concerned with fashion and pleasing themselves, than with valuing children for the children's own sake. However, we should remember that choosing a same race or same race-mix child is also designing the child that you will have. This is no less an exercise of parental preference than is the case of choosing a different race or race mix, or, for that matter, colour.

The best way both to avoid totalitarianism and to escape the possibility of racial (or gender) prejudice, either individual or social, dictating what sort of children people have, is to permit free parental choice in these matters. And to do so whether that choice is exercised by choice of procreational partner or by choice of gametes or embryo, or by genetic engineering. Such choices are, for the most part, likely to be as diverse as are the people making them.[21]

And, of course, the reference to genetic engineering in the above passage includes the idea of cloning. As I have suggested, I accept the high value that Putnam places on a willingness to accept diversity, and on families which embody this image of diversity (few and far between as they are). I also accept the connection of such images with the sorts of society they are likely to produce. Unlike Putnam, however, I conclude from this that we should accept diversity in family foundation including the use of cloning.

I am not, of course, insisting that it is clear that I am right and Putnam wrong about how to interpret the value of autonomy and diversity. What I do claim, however, is that the same, or at any rate similar, moral images are guiding both approaches. In such a case, tolerance of diversity seems to require that decent people be permitted to follow their own moral images in their own way, and while we may take different views about the use of cloning, neither should foreclose the other's options.

We will return to this point when we consider arguments for procreative autonomy in the last section of this chapter. The problems with Putnam's view are: how to make his image yield the conclusions he wants on the subject of cloning and not other equally compatible conclusions of the sort that I would draw; whether the moral image of the family is sufficient, and sufficiently determinate, to license the limitation on the autonomy of those who wish to use cloning; and whether his conclusions, and the way he interprets his moral image, license, and indeed encourage something else?

With respect to the last of these problems we should note that most societies accept and use images not very unlike those in the Nazi posters. An 'ideal' or idealized family, in adverts, for example, or in a 'sitcom', would be likely to display the same homogeneity. As would an 'ideal' or a 'typical' Jewish family or an ideal or typical African-American family. As I have suggested, we don't normally think that people founding such families are doing anything wrong when they decline to marry or procreate outside their ethnic group. If we take seriously Putnam's

Parthian shot, and support a 'right of each new-born child to be a *complete* surprise to its parents' (p. 13),[22] we should perhaps ask what would really surprise the parents of our imagined Jewish or African–American family? The answer would surely be that they would be genuinely surprised if their children turned out to be pure Aryan types straight from the Nazi posters! And the wicked pleasure that the reverse would cause, of a Nazi family giving birth to a Jewish or African (looking) child is surely justification in itself.[23] Should we use genetic manipulation to ensure that each family, like most good societies, is ethnically and culturally diverse, and that parents are always surprised by their offspring from the moment of birth, or even before?[24] The more serious point here is that parents will, of course, always and inevitably be surprised by their children. If they use cloning techniques for reproduction they may be less surprised by their children's physical appearance, but they will, for sure, be surprised by their children's dispositions, desires, traits, and so on—the more so if they expected them to be identical with those of the nucleus donor.[25]

Moreover, and this should not be forgotten, if parents have chosen their nucleus donor wisely, they may be less surprised by one thing: they should be less unpleasantly surprised by genetic diseases and defects, for they will not only know much about the nucleus donor, but will have had opportunities to carry out genetic tests before creating the clone. A moral image that limits these unpleasant and often catastrophic surprises is one we all should have constantly before us.

Autonomy Again

Axel Kahn produces a bizarre twist to the argument from autonomy which we should note before refocusing our discussion. Kahn defines autonomy as 'the indeterminability of the individual with respect to external human will' and identifies it as one of the components of human dignity. This is hopeless as a definition of autonomy; those in Persistent Vegetative State (PVS) and, indeed, all newborns would, on such a view, have to count as autonomous! However, Kahn then asserts:

The birth of an infant by asexual reproduction would lead to a new category of people whose bodily form and genetic make-up would be exactly as decided by other humans. This would lead to the establishment of an entirely new type of relationship between the 'created' and the 'creator' which has obvious implications for human dignity.

Federico Mayor recently endorsed[26] a version of this argument claiming: 'Cloning would remove the uniqueness that ensures no one has chosen and instrumentalized another person's identity.' Mayor has got his formulation back-to-front here. It is not the uniqueness that might ensure that no one has instrumentalized another, since the uniqueness does not precede the alleged instrumentalization. I imagine what Mayor believes is that it is the fact that we prevent instrumentalization that might help to ensure the uniqueness.

Either way, Kahn and Mayor are, I'm afraid, wrong on both counts. As Richard Gardner notes in his discussion essay on Ian Wilmut's chapter, even in the case of true cloning, mitochondrial genes, intrauterine influences, and subsequent nurture all affect bodily form and genetic identity. And Gardner's emphasis on the importance of nurture to the formation of personality in genetically 'identical' siblings I have, myself, repeatedly argued elsewhere. In short, it may be that 'manners maketh man' but genes most certainly do not.

Autonomy, as we know from monozygotic twins, is unaffected by close similarity of bodily form and matching genome. The 'indeterminability of the individual with respect to external human will' will remain unaffected by cloning. Where then are the obvious implications for human dignity? When Kahn asks: 'is Harris announcing the emergence of a revisionist tendency in bioethical thinking?' the answer must be that I am pleading for the emergence of 'bioethical *thinking*' as opposed to the empty rhetoric of invoking resonant principles with no conceivable or coherent application to the problem at hand.

Genetic Variability

It is often claimed that human cloning will reduce genetic variability with catastrophic results. The UNESCO approach to cloning for example refers to the preservation of the human genome as 'the common heritage of humanity'. Well, normal genetic reproduction does not preserve the human genome, rather it constantly varies it. If *preservation* were the issue, cloning on a universal scale would be the best way to achieve that (clearly dubious) objective. Only if all existing people were cloned would the human genome be 'preserved' intact. Hilary Putnam concludes the written version of his Amnesty lecture[27] with the assertion that: 'What I have been claiming is that the unpredictability and diversity of our progeny is an intrinsic value' (p. 12). We have noted in discussing Axel

Kahn's views that it is simply false that cloning could make children anything other than unpredictable. And, as to genetic diversity, cloning cannot be said to impact on the variability of the human genome, it merely repeats one infinitely small part of it, a part that is repeated at a natural rate of about 3.5 per thousand births.[28] Those who raise threats to the human genome as a fear in connection with human cloning owe us an explanation of how the human genome or genetic variability might be adversely affected.

It is a truism which has become a cliche that genotype is not phenotype, that genes influence, but do not determine, what traits people will have. Putnam has postulated fairly widespread availability of cloning. But given the cost and the foresight required (avoidance of sexual reproduction) it is never likely to be a standard method of reproduction anymore than is IVF. Natural monozygotic twins occur at a rate of 3.5 per thousand pregnancies. It is surely unlikely that even on Putnam's scenario, the cloning rate would match this. Would more than one in every 270 couples or individuals for that matter want, or be able to use, cloning technology? Suppose this rate were doubled or even tripled by cloning, would anyone even notice? If the rate of twinning is the issue we could of course regulate access to cloning technology rather than ban it altogether.

The resolution of the European Parliament takes a different tack. Having repeated the, now mandatory, waft in the direction of fundamental human rights and human dignity, it suggests that cloning violates the principle of equality, 'as it permits a eugenic and racist selection of the human race'. Well, so do prenatal and pre-implantation screening, not to mention egg donation, sperm donation, surrogacy, abortion, and human preference in choice of sexual partner. The fact that a technique could be abused does not constitute an argument against the technique, unless there is no prospect of preventing the abuse or wrongful use. To ban cloning on the grounds that it might be used for racist purposes is tantamount to saying that sexual intercourse should be prohibited because it permits the possibility of rape.

The second principle appealed to by the European Parliament states that 'each individual has a right to his or her own genetic identity'. Leaving aside the inevitable contribution of mitochondrial DNA,[29] we have seen that, as in the case of natural identical twins, genetic identity is not an essential component of personal identity,[30] nor is it necessary for 'individuality'. Moreover, unless genetic identity is required either for personal identity, or for individuality, it is not clear why there should be a right to such a thing. But if there is, what are we to do about the rights of identical twins?

Suppose there came into being a life-threatening (or even disabling) condition that affected pregnant women and that there was an effective treatment, the only side effect of which was that it caused the embryo to divide, resulting in twins. Would the existence of the supposed right conjured up by the European Parliament mean that the therapy should be outlawed? Suppose that an effective vaccine for HIV was developed which had the effect of doubling the natural twinning rate; would this be a violation of fundamental human rights? Are we to foreclose the possible benefits to be derived from human cloning on so flimsy a basis? We should recall that the natural occurrence of monozygotic (identical) twins is one in 270 pregnancies. This means that in the United Kingdom, with a population of about 58 million, over 200 thousand such pregnancies have occurred. How are we to regard human rights violations on such a grand scale?

Clearly the birth of Dolly and the possibility of human equivalents has left many people feeling not a little uneasy, if not positively queasy at the prospect. It is perhaps salutary to remember that there is no necessary connection between phenomena, attitudes, or actions that make us uneasy, or even those that disgust us, and those phenomena, attitudes, and actions that there are good reasons for judging unethical. Nor does it follow that those things we are confident are unethical must be prohibited by legislation or controlled by regulation. These are separate steps which require separate arguments.

Moral Nose

The idea that moral sentiments, or gut reactions, must play a crucial role in the determination of what is morally permissible is tenacious. This idea, originating with David Hume (who memorably remarked that morality is 'more properly felt than judg'd of'), has been influential in the work of a number of contemporary moral philosophers,[31] in particular, Mary Warnock has made it a central part of her own approach to these issues. Briefly the idea is:

If morality is to exist at all, either privately or publicly, there must be some things which, regardless of consequences should not be done, some barriers which should not be passed. What marks out these barriers is often a sense of outrage, if something is done; a feeling that to permit some practice would be indecent or part of the collapse of civilization.[32]

A recent, highly sophisticated, and thoroughly mischievous example in the context of cloning comes from Leon R. Kass. In a long discussion entitled 'The wisdom of repugnance' Kass tries hard and thoughtfully to make plausible the thesis that thoughtlessness is a virtue. 'We are repelled by the prospect of cloning human beings not because of the strangeness or novelty of the undertaking, but because we intuit and feel, immediately and without argument, the violation of things that we rightfully hold dear.'[33] The difficulty is, of course, to know when one's sense of outrage is evidence of something morally disturbing and when it is simply an expression of bare prejudice or something even more shameful. The English novelist George Orwell[34] once referred to this reliance on some innate sense of right and wrong as 'moral nose', as if one could simply sniff a situation and detect wickedness. The problem, as I have indicated, is that nasal reasoning is notoriously unreliable, and olfactory moral philosophy, its theoretical 'big brother', has done little to refine it or give it a respectable foundation. We should remember that in the recent past, among the many discreditable uses of so-called 'moral feelings', people have been disgusted by the sight of Jews, black people, and indeed women being treated as equals and mixing on terms of equality with others. In the absence of convincing arguments, we should be suspicious of accepting the conclusions of those who use nasal reasoning as the basis of their moral convictions.

In Kass's suggestion (he disarmingly admits revulsion 'is not an argument'), the give-away is in his use of the term 'rightfully'. How can we know that revulsion, however sincerely or vividly felt, is occasioned by the violation of things we rightfully hold dear unless we have a theory, or at least an argument, about which of the things we happen to hold dear we *rightfully* hold dear? The term 'rightfully' implies a judgement which confirms the respectability of the feelings. If it is simply one feeling confirming another, then we really are in the situation Wittgenstein lampooned as buying a second copy of the same newspaper to confirm the truth of what we read in the first.

We should perhaps also note for the record that cloning was not anticipated by the Deity in any of his (or her) manifestations on earth; nor in any of the extant holy books of the various religions. Ecclesiastical pronouncements on the issue cannot therefore be evidence of God's will on cloning, and must be examined on the merits of the evidence and argument that inform them, like the judgements or opinions of any other individuals.

Is it Possible to be Too Reasonable?

One small beacon of relative sanity has been the Shapiro Report, commissioned by President Clinton from the National Bioethics Advisory Commission.[35] Basically, the Shapiro Report has taken the line that cloning is insufficiently well developed or understood to be regarded as safe for use in humans as yet. Therefore, because it's unsafe it would be unethical to permit an unsafe procedure to be used. Shapiro therefore recommends '[F]ederal legislation should be enacted to prohibit anyone from attempting . . . to create a child through somatic cell nuclear transfer cloning.' However, there are two important features of the Shapiro Report that are worthy of attention. The first is that it explicitly emphasizes that '[I]t is critical . . . that such legislation include a sunset clause to ensure that Congress will review the issue after a specified time period (three to five years) in order to decide whether the prohibition continues to be needed.' Secondly, the report, almost uniquely among publicly sponsored reactions to Dolly, pays very detailed attention to the moral and policy arguments on both sides. It misses, however, one important opportunity. Having spelled out the arguments it makes no assessment of them; it merely recommends further widespread examination of the arguments in public fora. Since none of the arguments against cloning it reviews have any more plausibility than those we have just considered, there seems little to be gained by further reflection on them, or indeed iteration of them. We need authoritative analysis quite as much as authoritative consideration of the arguments.[36]

2. Does the Legislative Response to Dolly Constitute an Attack on Human Rights and Dignity?

The hysteria we have examined is, of course, fascinating. It tells us (we must presume) a great deal about the subconscious of the hysterics involved. It is also very disturbing because the individuals concerned occupy positions of power and influence world-wide and their gut reactions and prejudices have been translated into resolutions and regulations which are likely to prove seriously prejudicial to human welfare, and, more germane to our present concerns, to human rights.

Where nasal reasoning and olfactory moral philosophy generate bad arguments and these are deployed in the cause of limitations on human rights and freedoms this can itself constitute a grave attack on those freedoms. Few believe, or are prepared to admit, that human rights and freedoms may be restricted on a whim, or simply to assuage feelings of unease or even of revulsion, or to protect beliefs that are themselves expressions of prejudice or bigotry. One of the hallmarks of a moral position is the preparedness to deploy evidence and argument in its support. Where these are deployed they not only demonstrate the reasons for accepting the moral position and the adequacy of those reasons, but they also establish the position as morally respectable (if not correct) and its adherents as moral beings deserving of our respect. Where the arguments deployed in support of a moral position are inadequate or flawed then the position is left without justification and support and the conclusions which flow from it are unpersuasive. Where, however, the arguments are so thin, or implausible, as to be absurd, then not only is the position in support of which they are adduced unsupported, the moral integrity of those advancing it is undermined.

This, I judge to be the current position with respect to arguments against human cloning. Where, as is now the case, human cloning is the object of an effective worldwide moratorium backed in some cases by international resolutions and regulations, then those resolutions and regulations are without foundation or intellectual support.

My own view, which is for once widely shared, is that a fundamental principle of the morality of all democratic countries is that human liberty should not be abridged without good cause being shown. Now where the liberty in question is trivial or vexatious, or is itself morally dubious or even morally neutral, it is plausible to claim that no harm is thereby done. Particularly where popular support for the limitation of this supposed freedom can be demonstrated. However, where a case can be made to the effect that the freedom which has been abridged is not only not trivial, vexatious, or morally dubious but rather is itself the expression of or a dimension of something morally significant, then its abridgement becomes a serious matter.

The more so when we consider some of the likely candidates for human cloning. Everyone has in mind the Boys from Brazil scenario of maniacs choosing to clone Hitler's genome (but not of course Hitler, despite their hopes to the contrary) umpteen times in the jungles of South America. However, the fantasies of film-makers or megalomaniacs should not distort our response to real concerns and genuine need. We have examined one possible (but not very likely) compelling use of human cloning. Let's now look at a number of very likely cases which might be made for human cloning and test our intuitions about them before going on to examine the hard arguments for reproductive autonomy. We will assume in all cases that human cloning by nuclear substitution is as safe as was IVF when it was first performed.

Eight cases

1. A couple in which the male partner is infertile. They want a child genetically related to them both. Rather than opt for donated sperm they prefer to clone the male partner knowing that from him they will get 46 chromosomes, and that from the female partner, who supplies the egg, there will be mitochondrial DNA. Although in this case the male genetic contribution will be much the greater, both will feel, justifiably, that they have made a genetic contribution to their child. They argue that for

them, this is the only acceptable way of having children of 'their own'. In what way is their preference unethical or contrary to human rights or dignity?

2. A couple in which neither partner has usable gametes, although the woman could gestate. For the woman to bear the child she desires they would have to use either embryo donation or an egg cloned with the DNA of one of them. Again they argue that they want a child genetically related to one of them and that it's that or nothing. The mother in this case will have the satisfaction of knowing she has contributed not only her uterine environment but also nourishment, and will contribute subsequent nurture, and the father will have contributed his genes. In what way is their preference unethical or contrary to human rights or dignity?

3. A single woman wants a child. She prefers the idea of using all her own DNA to the idea of accepting 50 per cent from a stranger. What are the weighty ethical considerations that require her to be forced to accept DNA from a stranger and mother 'his child' rather than her own?

4. A couple have only one child and have been told they are unable to have further children. Their baby is dying. They want to de-nucleate one of her cells so that they can have another child of their own. In what way is their preference unethical or contrary to human rights or dignity?

5. One partner has a severe genetic disease. The couple want their own child and wish to use the other partner's genome. In what way is their preference unethical or contrary to human rights or dignity?

6. An adult seems to have genetic immunity to AIDS. Researchers wish to create multiple cloned embryos to

isolate the gene to see if it can be created artificially to permit a gene therapy for AIDS. In what way is their proposed course of action unethical or contrary to human rights or dignity?

7. As Michel Revel has pointed out,[37] cloning may help overcome present hazards of graft procedures. Embryonic cells could be taken from cloned embryos prior to implantation into the uterus, and cultured to form tissues of pancreatic cells to treat diabetes, or brain nerve cells could be genetically engineered to treat Parkinson's or other neurodegenerative diseases. In what way would this be unethical or contrary to human rights or dignity?

8. Jonathan Slack has recently pioneered headless frog embryos. This methodology could use cloned embryos to provide histocompatible[38] formed organs for transplant into the nucleus donor. In what way would this be unethical or contrary to human rights or dignity?[39]

3. Is the Creation of Human Dollys[40] Protected by Existing Conventions on and Assumptions Concerning Human Rights?

Procreative Autonomy

We have examined the arguments for and against permitting the cloning of human individuals. At the heart of these questions is the issue of whether or not people have rights to control their reproductive destiny and, so far as they can do so without violating the rights of others or threatening society, to choose their own procreative path. We have seen that it has been claimed that cloning violates principles of human dignity. We will conclude by briefly examining an approach which suggests rather that failing to permit cloning might violate principles of dignity.

The American philosopher and legal theorist Ronald Dworkin has outlined the arguments for a right to what he calls 'procreative autonomy' and has defined this right as the right of individuals 'to control their own role in procreation unless the state has a compelling reason for denying them that control'.[41] Arguably, freedom to clone one's own genes might also be defended as a dimension of procreative autonomy because so many people and agencies have been attracted by the idea of the special nature of genes and have linked the procreative imperative to the genetic imperative.

The right of procreative autonomy follows from any competent interpretation of the due process clause and of the Supreme Court's past decisions applying it . . . The First Amendment prohibits government from establishing any religion, and it guarantees all citizens free exercise of their own religion. The Fourteenth Amendment, which incorporates the First Amendment, imposes the same prohibition and same responsibility on states. These provisions also guarantee the right of procreative autonomy.[42]

The point is that the sorts of freedoms which freedom of religion guarantees, freedom to choose one's own way of life and live according to one's most deeply held beliefs, are also at the heart of procreative choices. And Dworkin concludes 'that no one may be prevented from influencing the shared moral environment, through his own private choices, tastes, opinions, and example, just because these tastes or opinions disgust those who have the power to shut him up or lock him up'.[43] Thus it may be that we should be prepared to accept both some degree of offence and some social disadvantages as a price we should be willing to pay in order to protect freedom of choice in matters of procreation and perhaps this applies to cloning as much as to more straightforward or usual procreative preferences.[44]

The nub of the argument is complex and abstract but it is worth stating at some length. I cannot improve upon Dworkin's formulation of it.

The right of procreative autonomy has an important place . . . in Western political culture more generally. The most important feature of that culture is a belief in individual human dignity: that people have the moral right—and the moral responsi- bility—to confront the most fundamental questions about the meaning and value of their own lives for themselves, answering to their own consciences and convictions . . . The principle of procreative autonomy, in a broad sense, is embedded in any genuinely democratic culture.[45]

The rationale that animated the principle of procreative autonomy was made the subject of a submission to the United States Court of Appeals by Ronald Dworkin and a group of other prominent philosophers. Their sub- mission was in a case concerning voluntary euthanasia and it is interesting because it cites a number of United States Supreme Court decisions and their rationale.

Certain decisions are momentous in their impact on the character of a person's life—decisions about religious faith, political and moral allegiance, marriage, procreation and death, for example. Such deeply personal decisions reflect controversial questions about how and why human life has value. In a free society, individuals must be allowed to make those decisions for themselves, out of their own faith, conscience and convictions. This Court has insisted, in a variety of contexts and circumstances, that this great freedom is among those protected by the Due Process Clause as essential to a community of 'ordered liberty.' *Palko* v. *Connecticut*, 302 US 319, 325 (1937).

In its recent decision in *Planned Parenthood* v. *Casey*, 505 US 833, 851 (1992), the Court offered a paradigmatic statement of that principle:

matters involving the most intimate and personal choices a person may make in a lifetime, choices central to a person's dignity and autonomy, are central to the liberty protected by the Fourteenth Amendment.

As the Court explained in *West Virginia State Board of Education* v. *Barnette*, 319 US 624, 642 (1943):

If there is any fixed star in our constitutional constellation, it is that no official . . . can prescribe what shall be orthodox in politics, nationalism, religion, or other matters of opinion or force citizens to confess by word or act their faith therein.

In note 1.1 to the Philosophers' Brief it is argued that:

Interpreting the religion clauses of the First Amendment, this Court has explained that 'the victory for freedom of thought recorded in our Bill of Rights recognizes that in the domain of conscience there is a moral power higher than the *State*.' *Girouard* v. *United States*, 328 US 61, 68 (1946).

And, in a number of Due Process cases, this Court has protected this conception of autonomy by carving out a sphere of personal family life that is immune from government intrusion. See e.g. *Cleveland Bd. of Educ.* v. *LeFleur*, 414 US 632, 639 (1974) ('This Court has long recognized that freedom of personal choice in matters of marriage and family life is one of the liberties protected by the Due Process Clause of the Fourteenth Amendment.'); *Eisenstadt* v. *Baird*, 405 US 438, 453 (1973) (recognizing right 'to be free from unwarranted governmental intrusion into matters so fundamentally affecting a person as the decision to bear and beget a child'); *Skinner* v. *Oklahoma*, 316 US 535, 541 (1942) (holding unconstitutional a state statute requiring the sterilization of individuals convicted of three offenses, in large part because the state's actions unwarrantedly intruded on marriage and procreation, 'one of the basic civil rights of man'); *Loving* v. *Virginia*, 388 US 1, 12 (1967) (striking down the criminal prohibition of interracial marriages as an infringement of the right to marry and holding that 'the freedom to marry has long been recognized as one of the vital personal rights essential to the orderly pursuit of happiness by free men').

These decisions recognize as constitutionally immune from state intrusion that realm in which individuals make 'intimate and personal' decisions that define the very character of their lives.[46]

In so far as decisions to reproduce in particular ways or even using particular technologies constitute decisions concerning central issues of value, then, arguably the freedom to make them is guaranteed by the constitution (written or not) of any democratic society, unless the state has a compelling reason for denying them that control. To establish such a compelling reason the state (or indeed a federation or union of states, like the European Union for example) would have to show that more was at stake than the fact that a majority found the ideas disturbing or even disgusting.

Federico Mayor has attempted to deny that fundamental principles about the liberty to reproduce and to found a family are at issue in the case of cloning. Referring to the *Universal Declaration of Human Rights* Mayor says: 'Article 16 defends a basic component of human life against

political prohibitions, not the "right" to use any technology to overcome physiological impediments to natural
reproduction'.[47] Of course Mayor is entitled to attempt to
interpret the expression of the basic value to be found in
Article 16. However, he neither explains, elaborates, nor
defends his interpretation, he merely stipulates it. The
right to found a family protected by Article 16 is at once
wider, and I believe more profound, than the right to
reproductive autonomy or liberty defended by Dworkin
and others. To argue for this is, I'm afraid, a task for
another occasion. I would, however, claim that Article 16,
if it is to be coherent at all, must include the right to
procreative autonomy as elaborated and defended above.
If this is right, a *prima facie* case has been made against
Mayor's, I believe excessively, restrictive interpretation of
so basic and fundamental a value.

As yet, in the case of human cloning, such compelling
reasons have not been produced. True, procreative
autonomy will not cover all proposed cases of human
cloning (for example, cases 7–9 above). Suggestions have
been made, but have not been sustained, that human
dignity may be compromised by the techniques of cloning.
Dworkin's arguments and the cases cited in the brief above
suggest that human dignity and, indeed, democratic con-
stitutions may be compromised by attempts to limit
procreative autonomy, at least where greater values cannot
be shown to be thereby threatened. I have argued that no
remotely plausible arguments exist as to how human
cloning might pose significant dangers or threats or that
it may compromise important human values. It has been
shown that there is a *prima facie* case for regarding human

93

cloning as a dimension of procreative autonomy that should not be lightly restricted.[48]

Cloning and Public Policy

Ruth Deech

Professor Harris's paper 'Clones, Genes, and Human Rights' is an important contribution to the debate over cloning. One cannot take issue with the strength of many of the philosophical arguments put forward by Harris. There are, however, wider societal and practical perspectives[1] on cloning which command our attention.

In 1991 the Human Fertilisation and Embryology Authority (HFEA) was established. This statutory body regulates the use and storage of human embryos and gametes outside the human body and gives advice to the public and to government on the ethical and scientific issues arising from progress in artificial reproductive technology. The Human Fertilisation and Embryology Act 1990[2] is the main reference point for the HFEA. Informing the 1990 Act were the recommendations of the Warnock Committee which, in its Report,[3] insisted that 'the embryo of the human species ought to have a special status',[4] and that this should be enshrined in legislation. In according the embryo a special status, the Committee did not seek to afford the human embryo the same status as a living child or adult. Instead, it aimed to prevent the frivolous or unnecessary use of human embryos. While the Committee did stipulate that some research on embryos should be permitted, it also said that it should only be carried out up to the fourteenth day of development,[5] and that all such research ought to be strictly controlled and monitored. These recommendations were included in the provisions of the Human Fertilisation and Embryology Act 1990, which states that any research performed on human embryos up to the fourteenth day of development must conform to certain restrictions and is not lawful unless it has been licensed by the HFEA.[6] In addition, the 1990 Act requires any clinician in deciding whether or not to offer infertility treatment to a patient to 'take account of the welfare of any child who may be born as a result of the treatment (including the need of that child for a father), and of any other child who may be affected by the birth'.[7]

In the United Kingdom, embryo creation which involved nuclear replacement technology would not, by law, be allowed to lead to any fetuses and/or babies being produced. Non-reproductive uses of this technology, as the law stands, may be the only possible ones. An example of one non-reproductive use of this technology is the creation of *in vitro* stem cells which might be caused to differentiate into specific cell types to provide insight into how regeneration of damaged human tissue might be induced without risk of rejection. This sort of potential application of nuclear replacement therapy does raise ethical concerns, but these concerns differ greatly from the ones raised by reproductive cloning.[8]

Professor Harris, drawing on the writing of Ronald Dworkin on procreative autonomy, suggests that prohibitions on reproductive cloning would violate an individual's freedom of reproductive choice. The modern emphasis on reproductive freedom is an understandable reaction to the restrictions that have been placed on sexual relations of many kinds by various regimes and cultures during this century. But can the idea of reproductive autonomy serve to justify non-interference by government with regard to human cloning? Harris thinks it can. I, myself, am not convinced. The first point to note is that reproductive autonomy cannot mean the freedom to breed as one likes. Many societies, including our own, have rules about, for example, incest, underage sex, bestiality, adultery, and abortion for good, principled reasons. Thus reproductive autonomy cannot mean that all individuals have the right to have a baby or to become a parent at will. For example, no man in the United Kingdom has this right. Men cannot prevent women from having abortions or from using contraception. Whenever a man wishes to have a child it must be with the full consent of the mother to conception, pregnancy, and child-raising. And, women along with men have no absolute right to remain nurturing parents—children may legally be taken into care when it is deemed by the court that their needs are not being met. Measured regulation of sexual relationships and parenthood is the very hallmark of a caring society, one which is concerned to protect the vulnerable.

Professor Harris's permissive stance on human cloning is much supported by the right to procreative freedom and therefore it is flawed. Harris glosses over two very important provisions of the Human Fertilisation and Embryology Act 1990, both of which are in place for sound ethical reasons. The first of these is the rule governing the use and creation of embryos. Few would disagree that it is unacceptable to clone humans if doing so would involve (as it most certainly would) wastage of huge numbers of embryos in the process. Second, the

1990 Act also requires that when assisted conception technology is used, due consideration must be given to the well-being of any resulting child. Scant attention is paid to this last issue.

Would cloned children be the butt of jibes and/or be discriminated against? Would they become a sub-caste who would have to keep to each other? Would they be exploited? Would they become media objects? (This is not an unlikely scenario given that Louise Brown, the first test-tube baby, is still in the media some 20 years after her conception.) Would a cloned child be subject to excessive control from its parents or from one parent, who may already be too dominating as evidenced by the choice of the cloning technique? Would cloning represent the desire of one parent to have a monopoly of control, annihilating any possibility of a second parent and family, and denying natural independence to the growing child? Would the sole parent be severely disappointed in the child if, as so often happens, s/he does not turn out as expected? I suggest that the aforementioned, very important concerns about child welfare merit close attention.

Harris's putative right to procreative autonomy cannot justify human cloning because this right would violate important provisions of the 1990 Act. Cloning a human by nuclear transfer technology would involve unlicensed and unlawful use of embryos, and it might very well compromise the psychological well-being of the cloned individual.

There is a further problem with human cloning that has been under-discussed. This is the question of social and genetic parentage. Cloning would involve an egg donor, a nucleus donor, and someone to gestate the 'clone'. Who would the real mother be? The woman who supplied the egg, the woman or man (!) who supplied the nucleus, or the woman who carried the embryo/fetus to term? The mother of the nucleus donor could be thought of as the grandmother of the clone, but she might also claim the title of mother to the clone because she has contributed 50 per cent of the clone's genetic material. If the nucleus comes from a woman's husband, is he then the mother? If the cell donor is deceased might s/he be the real mother? And, to insist on the previous point made, how would the cloned child feel about these novel, not to mention confusing, parental relations? What if the cloned child wishes to know about his/her genetic parent(s) later in life? Should this be permitted? Clearly, the issue of parentage presents a host of thorny problems, ones which are not readily resolvable by recourse to the so-called right to procreative autonomy.

Harris derides public opinion about cloning which, as he correctly points out, is an opinion shared by members of a number of reputable national and international legislative bodies and other organizations. Part of the HFEA's role is to reflect society's worries about such technology. Many people say that moral boundaries are being transgressed by the possibility of cloning in a way that they are not, say, by infertility treatments such as *in vitro* fertilization (IVF). The latter mimics natural reproduction, whereas cloning goes well beyond this. In particular, fears have been expressed that human cloning threatens our sense of our own identity and devalues human dignity. In the view of the public, there are some powers which perhaps no one should have. The public appears disinclined to contemplate the creation of babies by selective cloning and genetic manipulation. Underpinning this disinclination are many fears, some of which are irrational, some of which are perfectly warranted.

Of course, we want informed public reactions to guide policy, and we want to ensure proper freedom of scientific inquiry and research.[9] Harris calls the public reaction against cloning 'hysterical'. But, as my preceding remarks make clear, this characterization is misleading. It is perfectly reasonable of us to think that human reproductive cloning, even if used responsibly, violates the sensible rules that we have about the moral status of human embryos; the technique would undoubtedly involve unlawful creation and immense wastage of them. It is also reasonable to fear that a cloned child's well-being might well be threatened. And it is eminently reasonable to worry that cloning technology poses problems about parentage and that it might be abused in the absence of internationally enforceable regulations. Thus, while it is true that adverse public reaction to cloning has been strong, it is not as misplaced as Harris implies.

The refinement of the meaning and uses of cloning allow more precise consideration of the practical, social, and ethical issues involved. This is where philosophers such as John Harris have helped most—clarifying and defining the questions that individuals, society, and policy-makers need to address in light of Dolly. John Harris challenges some of the assumptions that underscore our reactions to the possibility of human cloning and, in so doing, he forces us to confront these issues and has moved the debate forward. Meanwhile the HFEA and others have a responsibility to ensure that the understandable fears of society at large are not forgotten.

Eugenics and Human Rights

Jonathan Glover

In this chapter I want to address a number of questions surrounding developments in genetics. The first of these will be the question of how far we should be guided by deep emotional reactions, sometimes called the 'yuk factor', in our deliberations about applications of genetic engineering in humans.

Eugenics and Listening to the Emotions

On one view, it is foolish not to listen to these responses. In relationships, much is lost by not listening to the emotions. Our intuitive responses to people often give clues to what they are like, clues which go beyond reasons we are conscious of, and which are sometimes backed up later by more public evidence. The unease aroused in us by some new use of biotechnology can be seen as a similar warning that deep values are at stake, values which we may not yet be able to articulate, but which may be important to our humanity. Perhaps a morality divorced from emotional responses would be a good one for rational Martians, but it would not do for humans. On a non-theological view, morality is only a codification of human values, with their rootedness in human emotional responses.

On the other hand, some kinds of revulsion, especially

against things which seem 'unnatural', are just prejudice, of the kind which did such harm to most gay people before our own time. This makes some philosophers impatient with the 'yuk factor'. They demand its replacement in serious discussion by arguments based on real risks and benefits.

I have some sympathy with the first view. When some new development arouses widespread disquiet, there is certainly a case for suspecting that something really deep may be at stake, something which may be missed by the most obvious kind of cost-benefit analysis. But this can be combined with sympathy for the second view. Perhaps we should start with the human emotional responses, but because what therapists like to call 'being in touch with your emotions' is not enough, we need to connect with the intellect too.

Immanuel Kant, speaking of the way the mind does not just passively receive sensory experiences but actively interprets them, said that we should interrogate nature 'not like a pupil but like a judge'. We need some such interrogation of our emotional responses if they are to be a basis for serious decisions about ethical questions. These responses are clues a detective will note. They are not the factor which in court would impress a good judge as settling everything.

Both the importance and the inconclusiveness of our first reactions come out in the case of 'eugenics'.

In our time, the word has a chilling ring. If a policy can be described as 'eugenic', that is enough for most people to rule it out at once. In one way this is utterly appropriate. The great twentieth-century eugenic Programme was that of the Nazis. The murder of millions of men, women, and children grew directly out of that Programme. The 'yuk' response to any revival of those ideas is something for which no one need apologize.

But it is also true that the word 'eugenic', like the word 'fascist', often is used as a denunciation with little precise content. Because the Nazi atrocities were of such enormity, one natural response is to find them undiscussable. But there are reasons for thinking that certain policies, already with us, which favour the birth of some kinds of people rather than others, may be defensible. The first emotional reaction against 'eugenics' is unhelpful when we need to discriminate between morally acceptable and unacceptable decisions.

Biological advances are making it increasingly possible in medical contexts to take decisions about the kinds of people who are born. For more than thirty years, ante-natal testing has made it possible to terminate a pregnancy when the fetus has some detectable serious disability. With *in vitro* fertilization, there is the possibility of transferring to the womb 'healthy' embryos rather than others. (Embryo transfer largely avoids entanglement with the abortion issue.) Sex selection to avoid sex-linked disorders is with us. Safe and reliable gene therapy may not be far off. And with us already is genetic counselling which may affect people's choice of spouse or partner. In Cyprus and in Sardinia, where thalassaemia is very common, genetic testing has moved back from pregnancy, first to pre-marital screening, and then to screening in schools.

There are also non-medical possibilities. Sex selection on grounds of parental preference is no more difficult than on medical grounds. Non-therapeutic genetic engineering may be less far off than we like to think.

This power to decide for and against certain human characteristics is terrifying because, as the Nazi case reminds us, in some hands it could be appallingly misused. Because this is terrifying, it is tempting to turn away from thinking about it. Those who oppose ante-natal screening claim it is eugenic and assume they have made their case. Those who are willing to consider the development of some of these technologies say, and probably think, that of course nothing eugenic will arise. The 1993 report of the Nuffield Council on Bioethics on 'Genetic Screening: Ethical Issues' devoted ten pages to issues about insurance and just over two to eugenics. Confronting future issues which seem dauntingly large, it is natural to turn to issues both current and manageable. But looking away is always a weak intellectual strategy. 'Yuk' may be the right first response to eugenics, but thought should surely be the next one.

Part of this should be some kind of analysis of what is meant by calling a policy 'eugenic' and what are the criticisms for which the word stands in. A first shot at characterizing a eugenic policy might be that it intends the birth of some kinds of people rather than others. The intention is crucial. Virtually all major social policies have a 'eugenic' impact in a weaker unintentional sense. The existence of the National Health Service, or changes in taxes and welfare benefits, will make a difference to who is born. These institutions and policies are not criticized as eugenic because the impact is not intended. But when, as at one time in Singapore, larger family allowances are given to university graduates, apparently with the aim of encouraging them to have more children, worries about eugenics seem very much to the point.

Human Rights

A word about the other half of the title of this chapter: human rights. I am a huge admirer of the human rights campaigns of Amnesty International, and of the splendid impartiality with which it conducts them. An organization which has criticized and in turn been attacked by such a variety of oppressive governments of so many political shades cannot but command respect. But I have to admit that in one way I am here under false pretences. Despite strongly supporting the campaigns in defence of political dissidents and against oppression, torture, and the death penalty, I am a bit uneasy about the concept of human rights.

That some things—freedom of speech, freedom of religion, freedom of movement, freedom from cruel or degrading treatment such as torture or execution—are so precious that they should be defended even if they get in the way of realizing desirable political or economic goals is an idea fully deserving of support. The unease is at a philosophical level. There seems so little that is clear and convincing about the criteria by which we recognize something as a right, or how we decide what to do in cases where apparent rights are in conflict. As Ronald Dworkin has said, rights are trumps. If your interests and mine conflict, some weighing up of their relative import-ance is called for, perhaps leading to a compromise. But if what my interest conflicts with is your right, my interest simply has to give way. This trumping power makes a right an immensely desirable asset in a political debate. There is a great temptation to back political positions with claims about rights.

The abortion debate is a case in point. One side has the unborn child's right to life. The other side has the woman's right to choose. Despite the frequency with which these rights are asserted, most campaigners have little helpful to say about how we know that one of these rights exists while the other does not. Or, if they both exist, about how to decide which has priority. An Amnesty Lecture is of course not the place to say that talk of human rights is 'nonsense on stilts'. In parts of the world with frequent floods, houses on stilts save people's lives. Those of us who support Amnesty International's work can hardly avoid the shorthand phrase: of course we support human rights. But some of us do so with the uneasy hope that the ground on which the stilts rest may be firmer than it looks.

In discussing various policies which may be eugenic, I will mention the question of whether they conflict with certain possible human rights. But, especially in the new ethical issues opened up by medical and biological advances, the talk of rights is and should be tentative.

Medical Decisions about Who is Born

The existence of ante-natal screening programmes for certain serious disorders, with the woman or couple being given the option of terminating the pregnancy if one is detected, makes a difference to how many people of different kinds are born. Some have seen this as an unacceptable case of eugenics. A few years ago, Dominic Lawson wrote an emotionally powerful article about the decision taken by his wife and himself to reject antenatal tests, and about the birth of their daughter with Down's syndrome. He likened the screening Programme to Nazi eugenics and said that a whole industry has been developed to make it increasingly improbable that children like his daughter will be allowed to live: 'This is nothing less than the state sponsored annihilation of viable, sentient fetuses.'

Other cases both reflect and suggest a different view. This was powerfully expressed by two parents whose daughter had died of the incurable genetic disease Dystrophic Epidermolysis Bullosa (EB). Her skin lacked essential fibres, so that contact with it caused large blisters which burst leaving raw open sores and terrible scars. Their daughter's condition extended to her digestive and respiratory tracts. The internal blistering and scarring caused her to have a painful life of only twelve weeks. There was a one-in-four probability that any future child the couple conceived would have EB. They decided to conceive again only because of the possibility of genetic testing followed by termination of the pregnancy. They wrote that 'such a decision is not taken lightly or easily . . . We have had to watch our first child die slowly and painfully and we could not contemplate having another child if there was a risk that it too would have to die in the same way.'

If to practise eugenics is to intend the birth of some kinds of people rather than others, these parents, in aiming at the birth of a baby without EB rather than one with the disorder, were making a eugenic choice. Many of you will feel, as I do, great sympathy with the parents of the girl with EB. Their concern is to avoid a short life of terrible suffering for a future baby. There are complications about this. If we think in terms of rights, for instance a right not to be born with such a terrible condition, there is a puzzle about whom the right belongs to. Can merely potential people, as yet unconceived, really possess rights? And if, as we perhaps want to say, it would be cruel to give a child such a life, there is a question about how anyone can be harmed by being born. Normally, to be harmed is to be made worse off than you otherwise would be. But can we really make comparisons with the 'state' of being unconceived?

These conceptual complications have afforded much embarrassment to courts when people have brought actions for 'wrongful life', and have given philosophers new scope for the enjoyable exercise of ingenuity. As a philosopher, I love these complications. But they should not be allowed to obscure the essential validity of the parents' concern. If they did have another child with the same condition, that child would then have a life that might be worse than no life at all. The couple seeking to avoid this was humane and in no way confused.

But screening for Down's syndrome cannot be justified in the same way. Some children with Down's syndrome have related medical complications which make their lives very difficult. But the great majority of people with this condition have lives which are far from not worth living. So, in cases of Down's syndrome, attempts to justify termination of a pregnancy by appealing to the interests of the potential child are unconvincing. For those who believe that fetuses have an absolute right to life, these screening programmes will always be unjustifiable. But many of us are prepared to accept that a woman's choice can sometimes justify abortion. The special problems of caring for a child with a serious disability and the relative lack of supporting help for parents who do so may give potential parents a reason for choosing to terminate the pregnancy. Accepting that they have a right to do so fits well with a pro-choice stand on abortion.

Those who liken the screening Programme for Down's syndrome to Nazi eugenics overlook the difference between a policy based on respect for parental autonomy and a state policy of 'tidying up' the world by deliberately eliminating whole groups who did not fit the state blue-print of the ideal person. But, even for those of us inclined to see parental autonomy as a reason for accepting termin-ation of pregnancy in such cases as Down's syndrome, there is also a possible human cost to the screening Pro-gramme which should not be ignored. The screening Programme may pose a real threat to equality of respect for those now alive with conditions such as Down's syn-drome. It cannot do much for someone's self-esteem to have the thought that many parents are choosing abortion rather than have a child with the same condition they have.

If the screening Programme is to be morally acceptable, it must be accompanied by a serious campaign to change attitudes towards people with disabilities, and to protect their status in society. To some, this will not be enough to protect equality of respect. If they are right, there is a conflict between parental autonomy and equality of respect, which resembles the other conflict with any right to life the fetus may have. I am not here trying to resolve this screening issue, but to tease out some of the possibly conflicting values which are relevant to any humane debate on it.

Related issues are raised by the possibility of treating genetic disorders by developing safe and reliable forms of gene therapy. One argument for the acceptability of this points out that we intervene medically to deal with the consequences of the genetic defect. Gene therapy is only an intervention earlier in the causal chain. And gene therapy may seem to avoid 'eugenic' problems. It is a matter, not of terminating a pregnancy so that a different embryo or fetus can later come to term instead, but of curing the disorder in the existing embryo or fetus. One anti-eugenic view can be expressed by saying that we should prefer the slogan 'make people healthy' to 'make healthy people'. Gene therapy largely fits this. But there are complications.

One complication has to do with the contribution of genes to a person's identity. Perhaps replacing a gene, which contributes to a disorder, is little threat to the identity of the person. But massive genetic changes could blur the boundary between changing the characteristics of one person and replacing one person by another. The other complication is germ-line gene therapy. This means that the genetic change is passed down to future generations. Germ-line gene therapy can be seen as not just curing a disorder in one person but as changing the gene-pool.

The Threat to Human Rights: The Nazi Case

To look at our own emotional responses, 'not like a pupil but like a judge', the feelings of horror and revulsion which many people have when thinking about these issues are linked to the Nazi episode. It is necessary to look hard at what made the eugenics of the Nazis so terrible. The Nazi eugenic policy had three features which mark it off sharply from current medical debates. The Nazis had a blueprint of the most desirable type of person. They believed in Social Darwinism and 'racial self-defence'. And they were indifferent to the autonomy or interests of particular individuals.

Blueprint of the Best Type of Person

There was the idea that only the best people should be encouraged to procreate. The *Lebensborn* Programme was set up for this. There was a supposedly scientific basis for belief in a distinct 'Aryan' type of person. In practice, the criteria for choosing the 'best' people were very crude and mainly physical. (Roughly, the best people had to look not at all like Hitler.) The Nazis stole Polish children who seemed to have the right appearance. It is hard to believe that, even on Nazi assumptions, this part of the project can have seemed to have any very secure scientific basis.

The other part of the Programme was that some should be discouraged from having children, or even prevented from doing so. In 1934 one of the fathers of Nazi eugenics, Professor Fritz Lenz, said, 'As things are now, it is only a minority of our fellow citizens who are so endowed that their unrestricted procreation is good for the race.'[1] In 1923 Lenz, together with his colleagues Erwin Baur and Dr Eugen Fischer, wrote a textbook, *Outline of Human Genetics and Racial Hygiene*, said to have been read by Hitler, and whose ideas find echoes in *Mein Kampf*.[2] These ideas influenced the Sterilization Law brought in when Hitler came to power in 1933. This made sterilization compulsory for people with conditions including schizophrenia, manic depression, and alcoholism.

Social Darwinism and 'Racial Self-Defence'

The Nazi theorists were concerned with Darwinian natural selection. They were afraid that the 'natural' selective pressures, which had functioned to ensure the survival of healthy and strong human beings, no longer functioned in modern society. Because of such things as medical care, and support for the disabled, people who in tougher times would have died, were surviving to pass on their genes. Social Darwinism saw the evolutionary struggle as taking place not just between individuals, but also between groups. Nazism emerged against this background of belief in life as a ruthless group struggle for survival. According to Social Darwinism, victory goes to the strong, the tough, and the hard, rather than to those who are gentle and co-operative. The Nazis took this up. They extolled struggle and the survival of the fittest.

Hitler was a Social Darwinist. One day at lunch he said, 'As in everything, nature is the best instructor, even as regards selection. One couldn't imagine a better activity on nature's part than that which consists in deciding the supremacy of one creature over another by means of a constant struggle.'[3] Social Darwinism was part of what led the Nazis to abandon traditional moral restraints. One Nazi physician, Dr Arthur Guett, said: 'The ill-conceived "love of thy neighbour" has to disappear... It is the supreme duty of the ... state to grant life and livelihood only to the healthy and hereditarily sound portion of the population in order to secure ... a hereditarily sound and racially pure people for all eternity.'[4] The ideology was also one of racial purity. There was the idea that genetic mixing with other races lowered the quality of people.

In 1933 Dr Eugen Fischer was made the new Rector of Berlin University. In his Rectoral Address he said:

The new leadership, having only just taken over the reins of power, is deliberately and forcefully intervening in the course of history and in the life of the nation, precisely when this intervention is most urgently, most decisively, and most immediately needed . . . This intervention can be characterized as a biological population policy, biological in this context signifying the safeguarding by the state of our hereditary endowment and our race.

Fischer in 1939 extended this line of thinking specifically to the Jews. He said: 'When a people wants to preserve its own nature it must reject alien racial elements. And when these have already insinuated themselves it must suppress them and eliminate them. This is self-defence.'[5] Some appalling imagery was used to give racism a biological justification. Appalling images likened Jews to vermin, or to dirt and disease. When all Jews were removed from an area, it was called *Judenrein*—clean of Jews. Hans Frank, talking about the decline of a typhus epidemic, said that the removal of what he called 'the Jewish element' had contributed to better health in Europe. The Foreign Office Press Chief Schmidt said that the Jewish question was, as he put it, 'a question of political hygiene'.[6]

This kind of medical analogy was important in Nazi thinking. Hitler said:

The discovery of the Jewish virus is one of the greatest revolutions that have taken place in the world. The battle in which we are engaged today is of the same sort as the battle waged during the last century by Pasteur and Koch. How many diseases have their origin in the Jewish virus! . . . We shall regain our health only by eliminating the Jew.[7]

Unimportance of the Individual

The Nazi ideology was not one of the importance of the individual. There was a conception of the pure race and of the biologically desirable human being. Reproductive freedom and individual lives were to be sacrificed to these abstractions. Compassion for an individual victim was a weakness to overcome. Having been taught this, people working in the Nazi eugenic and 'euthanasia' programmes felt guilty about feelings of compassion. One Nazi doctor asked to kill psychiatric patients as part of the 'euthanasia' programme wrote to the director of the asylum. He explained his reluctance to take part in murdering the children there:

The new measures are so convincing that I had hoped to be able to discard all personal considerations . . . I cannot help stating that I am temperamentally not fitted for this. As eager as I often am to correct the natural course of events, it is just as repugnant to me to do so systematically, after cold blooded consideration, according to the objective principles of science, without being affected by a doctor's feeling for his patient . . . I feel emotionally tied to the children as their medical guardian, and I think this emotional contact is not necessarily a weakness from the point of view of a National Socialist doctor. I prefer to see clearly and to recognize that I am too gentle for this work than to disappoint you later.[8]

This apology for his concern for his patients, his emotional tie to these children, as 'not necessarily a weakness in a

National Socialist doctor', shows how deeply ingrained this ideology was.

Another medical model had great influence on the Nazis. It is an appalling medical model: the idea that in treating people who are 'racially inferior', you are like the doctor who is dealing with a diseased organ in an otherwise healthy body. This analogy was put forward in a paper in 1940 by Konrad Lorenz, the very distinguished ethologist, now remembered for his work on aggression, and whose books on animals had an enormous charm. Lorenz wrote this:

There is a certain similarity between the measures which need to be taken when we draw a broad biological analogy between bodies and malignant tumours, on the one hand, and a nation and individuals within it who have become asocial because of their defective constitution, on the other hand . . . Fortunately, the elimination of such elements is easier for the public health position and less dangerous for the supra-individual organism, than such an operation by a surgeon would be for the individual organism.[9]

The influence in practice of this thinking can be seen very clearly in Robert Jay Lifton's book on the Nazi doctors. He quotes a doctor called Fritz Klein. Dr Klein was asked how he would reconcile the appalling medical experiments he carried out in Auschwitz with his oath as a doctor. He replied: 'Of course I am a doctor and I want to preserve life. And out of respect for human life, I would remove a gangrenous appendix from a diseased body. The Jew is the gangrenous appendix in the body of mankind.'[10]

Human Values and the Medical Decisions

The values behind Nazi eugenics have only to be stated for the contrast with current medical and ethical debates to be clear. But it is worth drawing together some of the values which together add up to a strong defence against the kinds of things the Nazis did. Take the values which figure in the debate about ante-natal screening: the right to life, compassion for an individual child who may suffer, respect for the autonomy of the potential parents, equality of respect for people with disabilities. Each of these concerns would be utterly alien to Nazism, which would see them all as impediments to success in the racial struggle.

With Nazism in mind, it may be worth spelling out that, while it may be acceptable for parents to take some decisions about the kinds of children they will have, decisions of this sort by the state are utterly repugnant. Parental autonomy is not hard to discriminate from state eugenics. And, in a similar way, the concern to avoid harm to a particular future child is not the same as wanting to make improvements to the population or the gene-pool.

It is worth mentioning a way in which the rough working account of 'eugenics', as aiming at the birth of some kinds of people rather than others, might be narrowed down to exclude the parents of the girl with EB. If we want to keep the word for the really controversial cases, we can stipulate that a policy is eugenic only if it has the aim of influencing the composition of the population. It is easier to be against eugenics if it is defined so as to exclude cases like the parents of the girl with EB.

There are some cases which hover on the border of

eugenics, defined in this way. Thalassaemia, a recessive genetic disorder which requires invasive treatment and which drastically shortens life, is common in some Mediterranean countries. In Cyprus and in Sardinia schoolchildren are screened for thalassaemia. With this knowledge, people are able to choose partners so that they will not have children with the disease. Those carriers who deliberately seek non-carrier partners are typically not concerned with the gene-pool, but with the nature of their own future family. But those who set up the Programme might have had 'eugenic' thoughts about the gene-pool among their concerns. If so the policy both is and is not eugenic. Those of us with the anti-Nazi values outlined have reasons to support it. There is the autonomy of the parents and the harm thalassaemia does to those who have it. But there is also the thought that here is a slippery slope worth watching carefully.

Social Darwinism and Sociobiology

Next, I will make some comments on the intellectual source of much of what was bad in Nazi eugenics: Social Darwinism. Because of its role in Nazism, Social Darwinism is—almost universally and totally justifiably—discredited. But, current biological and medical developments give a reason to try to bring into focus some of the issues it raises. Another reason comes from the modern debate about sociobiology.

One mistake made by Social Darwinism was the assumption that groups such as nations or races are the unit of evolutionary competition. Modern evolutionary theory tends to see genes as the Darwinian units, with groups having at best a secondary instrumental role. Social Darwinists have a problem explaining which groups are the ones to consider. Although a British export, the theory in the early twentieth century was most influential in Germany. It was assumed that Germany was engaged in mortal biological combat with Britain and other countries. To see this as a biologically determined struggle showed a strange historical myopia. Less than a century before, Germany was not a state. Earlier Social Darwinists might have been more concerned with the evolutionary struggle between Prussia and Austria or Bavaria.

Modern sociobiology does provide a possible basis for giving groups a role in the 'survival of the fittest'. For the survival of my genes, saving the lives of several of my family may be more important than saving me. W. D. Hamilton's concept of 'inclusive fitness' combines personal fitness with this kinship component. Shaw and Wong have argued, in their book on the genetic seeds of warfare, that inclusive fitness may explain group conflict. A disposition to protect genetically related people from attacks by others could help gene survival. Fear and hostility directed towards other groups may have survival value.

Supporters of this theory need to explain how it applies to nations and other groups much larger than genetically related kin groups. More importantly, the theory suffers from a general problem of sociobiology. What are the constraints on such an explanation? Almost anything can be argued to be just what you would have expected to emerge from evolutionary competition. If most groups felt friendly towards each other, this could have been explained in terms of the evolutionary advantages of reciprocal altruism. There is a need for methods of testing, of distinguishing what is scientifically supported from what is still only a plausible story.

Another mistake of Social Darwinism was the assumption that ruthless struggle is the only survival strategy. Modern sociobiology suggests a variety of strategies for gene survival, including co-operative ones. 'Reciprocal altruism' may be to the benefit of the genes of both parties. Even modern sociobiology does not seem to leave much room for the kind of non-reciprocal altruism which some of us think is an important part of ethics. Sending money to help famine victims in a distant part of the world is hard to defend in terms of any likely reciprocal help. Some sociobiologists see that the constraints of maximizing gene survival do not easily accommodate some of our values. Richard Dawkins, expressing the hope that man has the capacity for disinterested altruism, says that we have the self-understanding which enables us to defy the selfish genes we were born with: 'we have the power to turn against our creators. We, alone on earth, can rebel against the tyranny of the selfish replicators.'[11]

One question about this is how much room Darwinian theory allows for a person or a group or a species doing this. Will those who do this not lose out to more ruthless competitors? And, if we all join together to do it, may there not later on be some mutation producing individuals who are more ruthless and so in the end supplant us?

Some comfort is to be drawn from game theory. Prisoners' Dilemma shows that individual pursuit of self-interest can be self-defeating, losing out to a more co-operative spirit. Much has been made of the encouraging results of Robert Axelrod, who showed that in tournaments of repeated Prisoners' Dilemma games, the conditionally co-operative strategy Tit for Tat defeated its rivals. There are two limitations to the encouragement to be drawn from this result. One is that, in competitive games, only one carefully structured class of pay-offs counts as Prisoners' Dilemma. Where the pay-offs are different, ruthless self-interest may not be self-defeating. The other limitation is made clear by Matt Ridley in *The Origins of Virtue*. Even within Prisoners' Dilemma, Axelrod's is only one kind of tournament. Later versions incorporate more features of the world and of human psychology. One version includes the fact that players may make mistakes. Another includes the human tendency to repeat winning strategies and to change losing ones. In some of these modified tournaments, Tit for Tat is not the winner.

The lesson of Prisoners' Dilemma is not that co-operation is always to a person's advantage. It is that under some conditions co-operation is more successful than selfishness. Those of us who want to see co-operation rather than conflict have a reason to try to rig the social rewards and benefits so that co-operation becomes a winning strategy.

In modern sociobiology there is a window of opportunity for something like a revived Social Darwinism. But, with luck, it is a window which can be closed. The window of opportunity is that, under some conditions, ruthless selfishness by an individual or by a group might be a winning strategy. Even at the abstract level of game theory, there is no guarantee against nice guys coming last. Everything depends on the structure of the contest.

The chance of closing the window is that we can see this. If we choose, we can set about rigging the structure of society and of the world to favour altruism and co-operation. Sometimes ruthless aggression pays off. It did for some in what was once Yugoslavia. But perhaps we can learn from this and, for instance, give enough (properly supervised) power to the United Nations to make this kind of aggressive behaviour a much poorer bet.

Some of the problems of survival seemed to the Social Darwinists to admit only of eugenic solutions. Without the constricting assumptions of that theory, these same problems look possibly amenable to social and political solutions. Remembering where those eugenic solutions led, we may feel some relief that this is so.

Human Nature, Emotions, and the Conversation of Mankind

As the work of Amnesty International testifies, we live in a time when human beings still engage in war, oppression, and cruelty on a terrible scale. The Hobbesian solution, of rigging the terms of the game so that these activities do not pay, is what we now need to aim for. Some optimists about eugenics would see this as a second best. Perhaps the dispositions which lead to war and cruelty are deeply programmed inside us. Perhaps they are in part genetically programmed. The optimists might hope to go beyond the Hobbesian strategy of containing our dangerous dispositions, and to change those dispositions themselves by eugenic means. This thought has real power. But, for such a eugenic strategy to work, hugely invasive genetic interventions would have to be applied to the whole human race. It seems fraught with danger and very impractical. The Hobbesian strategy is preferable.

But I share the sense that a strategy of containment by

threats and penalties is second best. It would be better if our psychology had less need of being contained. When we return from the abstractions of game theory to real people, there are possibilities in the space between Hobbesian containment and eugenics. There is the possibility that, through understanding more about the parts of our psychology that lead to atrocity and war, we may gradually recreate ourselves through cultural evolution.

I want to go back to our emotional responses. Some of our emotional responses are very dangerous: tribal, patriotic, or nationalistic emotions, the excitement at the thought of war, sadistic enjoyment of cruelty, the desire for revenge. We also have emotional responses which restrain us from atrocities: we sympathize with the distress of others, we have the capacity to be disgusted when people are humiliated or treated with cruelty. Wars and atrocities take place when, in some people, the darker emotions dominate and the countervailing human responses are absent or too weak to restrain them. Sometimes the human responses break through. George Orwell, in Spain to fight fascists, looked across at the fascist lines and saw an enemy soldier he could shoot. But the enemy soldier had been relieving himself and was running while holding up his trousers. Orwell did not shoot, because of the bit about the trousers: 'a man who is holding up his trousers isn't a "Fascist", he is visibly a fellow creature, similar to yourself, and you don't feel like shooting at him.'

The process of interrogating our emotions not like a pupil but like a judge is more complex than I have so far said. Atrocities and wars take place partly because of the times when the kind of human response Orwell describes is absent or can be turned off. Torturers demonize and dehumanize their victims to inhibit their own restraining human responses. People too easily support wars because future death and injury to unknown people who live far away do not draw an emotional response.

I am struck by the contrast between how people think about life and death decisions in personal or medical contexts and how they think about them in the context of going to war. When a baby is born with a severe disability which allows only a poor hope of a good life, the family and the doctors discuss what efforts if any should be made to keep the baby alive with great involvement and seriousness. Hardly anyone approaches the question of whether to support a war which may cost thousands of lives with anything like this seriousness. There is something dangerously missing in us.

But the question is whether things have to stay like this. We are rational and can see the inconsistencies in our first emotional responses and of trying to work out something more coherent. We are capable of seeing that at times we have something dangerously missing. And because of this, there is the possibility to some extent of correcting for our limitations. We can learn to use our imagination and to draw on our past experience. Our emotional make-up is not something fixed, but something which evolves, partly through our participation in what Michael Oakeshott called 'the conversation of mankind'.

The eugenic improvement of human nature is a project surrounded by dangers and nightmares. But the changing of human nature is not a project we need give up on. It is just that it is more in accord with our most humane values to opt for the slower and less dramatic strategy: changing ourselves and each other through the conversation of mankind.

Eugenics and Genetic Manipulation

Alan Ryan

I shall begin this commentary on Professor Glover's lecture
with two thoughts about eugenics in general, then turn
to the question of how much genetic manipulation we
should welcome, in the hope of saying a little more about
the so-called 'yuk factor' that Professor Glover mentions.
The first thing I would wish to emphasize about eugenics
is its origins in broadly progressive, and often in left-wing,
thinking. Today when we think of eugenics, we think
inevitably of the Nazis. We ought not to do so. Nor ought
we to think of the Nazis when we discover that writers
at the turn of the century—Bertrand Russell was among
them—feared what they called 'the degeneration of the
race'. We ought certainly to flinch at the lack of humanity
with which Russell, like H. G. Wells, contemplated the
disappearance of the aboriginal peoples of underdeveloped
countries: something they thought was sad but inevitable.
But the nonsense of Aryan supremacy was not what fuelled
their views.

Rather, it was revulsion at the effect of urban life on
the poorest inhabitants of late Victorian cities. In the
absence of the sort of understanding of genetics that came
fifty years later, it was all too easy to think that social
isolation and surroundings that were indisputably bad for

both body and mind, a setting which offered so many perverse incentives—rewarding casual relationships, penalizing careful child-rearing, making it irrational to forgo present pleasure for future well-being—would lead to the creation of a hereditary underclass. It is easy to see the attractions in a programme of positive eugenics, and it is not surprising that both in Britain and the United States the advocates of family planning were interested not only in happier marriages but also in those marriages producing healthier and fitter offspring. What is less easy to understand is the ease with which these benign motives underpinned the negative programme which envisaged a programme of sterilizing the mentally ill or 'feeble-minded'. Given the way in which screening for feeble-mindedness operated in the first two decades of the century—the statistics for the wretched foreigners who streamed through Ellis Island in the first fifteen years of the century particularly afford food for thought—any such programme was sure to be arbitrary, brutal, and inhumane.

Since such programmes were created, not by Nazis, but by people who on the whole tried to behave humanely and intelligently in the rest of their lives, one wonders why it was so hard for them to understand what they were doing. (I should say in passing that one might make much the same remark about both the British and the United States immigration systems today, where politicians of at least modestly humane views preside over systems of predictable and persistent inhumanity.) The sterilization statutes that remain to this day on the statute books of New Jersey and several other states were put there by progressives, not by antediluvian bigots. One might speculatively wonder whether one sort of progressivism contains within it an irrational fear of disorder, and thus also a fear not so much of miscegenation as of the presence of people whom they find odd and unfathomable—such as newly arrived immigrants from Eastern Europe and the sexually and morally dissident, along with the genuinely intellectually impaired.

I throw out these speculations to raise a question. Is well-meant intervention of the sort that genetic engineering permits perhaps the expression of an irrational desire to make the world a tidier, if also a happier, place? It might, on the other hand, have more to do with a desire to apply simple principles to a complicated world. Contemporary scientists sometimes seem oddly blind to the fact that other people do not share the rather simple utilitarian vision that they themselves accept. Asked whether it is a good idea for parents to be able to abort fetuses that will certainly suffer from muscular dystrophy if they come to term, they reply that it is, and most people agree. Press the discussion a little further, and they can find themselves saying that if there were a 'gay gene' and our offspring carried it, parents should have the choice to abort such a fetus, too. When I call this a 'rather simple utilitarianism', I skate over the discrepancy between the view that locates all important values in choice, and the view that holds that choice is only an imperfect (though perhaps the most reliable) guide to the chooser's utility.

Professor Glover starts, naturally enough, from the ill-repute into which eugenics has fallen, and holds the Nazi atrocities firmly at bay. He is, on the other hand, not inhospitable to the idea of modest measures of genetic engineering or, as one might say, modest measures of tailoring the nature of our offspring to our wishes as parents. I don't in any very deep way demur from any of this, but I should like to push the argument a little further. We start, I imagine, from a consensus on the thought that when we know how to eliminate some genetically based diseases without eliminating the fetuses that carry them, we should apply the technology to eliminate those diseases. In general, this consorts with our usual sentiment in favour of reducing misery, or alleviating harm, before we move on to change things for the better. We face none of the metaphysical puzzles that can plague discussions of *who* benefits when we, say, abort one damaged fetus, but immediately conceive another, numerically different but otherwise identical in all save the defective feature. In this case, it is quite clear that the child born as a result of the intervention at the fetal stage is the very same child as the child who would have been born damaged but for the intervention.

The case that pulls our intuitive responses in several different directions is, I think, the case of positive engineering. This is the case that discussions of the possibility of cloning humans have always focused on. There has already been a good deal of astute deflation of the more extravagant claims made by anxious or enthusiastic observers. The idea that we shall all engage in the elaborate, expensive, time-consuming, and intrinsically uninteresting processes by which a cloned offspring is produced hardly withstands five minutes' reflection on the pleasures of sexual intercourse. Why would we wish to produce a dubious copy of one parent when we can produce a more interesting non-copy of two, and by so much more agreeable a process? Moreover, the attractions of cloning oneself are much overrated. Most of us would quite like a version of ourselves that could go through life without making the same mistakes as we have done, but at the same time doing and enjoying the better things we have done and enjoyed. But the terms of the discussion rule that out pretty swiftly. Either the cloned child is just like us, in which case it will be a frightful disappointment to any but the most egregiously self-satisfied parent; or it is not just like us, in which case we might as well have children in the usual way.

This is, perhaps, too easy. The technology of cloning is still so far away from being able to support such fantasies that we might do better not to entertain them at all. Let us back away a little and ask a slightly different question. If we think that the limits of eugenic thinking ought to be set in rather the way ordinary public health thinking is set—that is, we are happy to have vaccination and inoculation programmes to eliminate smallpox, diphtheria, measles, and the like, do not mind the fluoridation of water to improve dental health, but would flinch from compulsory exercise or the legal regulation of diet—we may ask where the cut-off point comes with tinkering with our progeny's genetic endowment. The thought that 'defensive' measures come morally more readily than 'aggressive' or even 'progressive' measures is not hard to defend. The further we get away from a defensive baseline, the more room we give to unexpected and distasteful or distressing side-effects. Some of us, for instance, have doubts about the impact of new technology that allows us to save premature infants with extremely low birth-weights and underdeveloped lungs and other vital organs. Part of the unease is about the irrationality of spending so much money on heroic measures when there is a shortage of money for less heroic but more effective medical services. That is a fairly straightforward utilitarian anxiety.

But some of it is about the danger that we shall organize our dealings with the physical world in a way that is eventually very bad for us. The 'badness' for us, however, is rather hard to explicate, and it brings us back to the 'yuk factor'. One thing we might feel about heroic but misguided attempts to rescue very premature babies—perhaps in something of the same way as we feel about heroic but misplaced attempts to resuscitate elderly patients—is that there is another way of looking at the matter that has been crowded out by technique. A very premature baby that dies can be mourned, just as an elderly parent who dies can be mourned, but both can be mourned without our wishing to revisit the events leading up to their death with a view to altering them. One might think that a proper attitude to the death of a premature baby is to acknowledge that some human beings never quite make it into the world, just as all of us will eventually make our way out of it. Relentlessly trying to push back the boundaries of the possible may not only have diminishing marginal returns, but even negative ones. The premature baby 'loses' its life in a very attenuated sense; it is not cut off in mid-career so much as it fails to get to the starting gate.

To say so sounds like a callous unconcern with the disappointment, particularly of the mother frustrated of her chief concern at this moment in her life. Yet, it is not merely callous to wonder whether mothers should be so encouraged. When the heroically rescued child turns out to be brain damaged or otherwise unhealthy for the rest of a rather short life, as is all too often the case, we might wonder whether the child's mother ought to have been encouraged to take a different view. Similarly, where the families of elderly patients preserve the bare existence of the dying, it is not merely callous to suggest that they might do better to let go sooner rather than later.

To put it another way, the problem with following Francis Bacon down the track of achieving all things possible is that some of these things are rather dubiously for the relief of man's estate. And some of them begin to change our understanding of what we are doing in such a way that the emotions that naturally and readily attach to an activity described in one fashion will not so readily and naturally attach to an activity better described in other ways. One might think that the ability to, say, affect the hair colour of one's offspring was harmless enough; even then, the sensible parent might pause to wonder whether there was any guarantee that their child would share the parent's taste, and whether in a litigious age they mightn't have exposed themselves to lawsuits in thirty years' time. But the thought that hair colour is 'a matter of taste', so that the harmless vanity of parents can decently be appeased, looks persuasive enough.

Then, however, we might press the thought a little further, so that all sorts of other features of our future progeny can be controlled. At what point do we start thinking of having a baby as more like buying a toy from a mail-order catalogue than what it has hitherto been? For if it were like that, the humanitarian arguments in favour of elaborate intervention become less plausible. Someone who said that she was desperate to purchase a 'cabbage patch' doll would be thought to be a case for treatment rather than assistance. It is precisely because babies are not like toys that we rightly mind so much about having them; but if we push our technical facility to the point where we can choose what sort of baby to have in this consumerist fashion, would we not have undermined the whole purpose of doing so in the first place?

Silver Spoons and Golden Genes: Talent Differentials and Distributive Justice

Hillel Steiner

Introduction

To set the context for the argument I want to advance, it may be helpful to say a few general words about both distributive justice and talents. Distributive justice is concerned with judgements about who should get what and who should give what. But it is not concerned with all such judgements. Rather, only those gettings and givings which are thought to be justifiably enforceable fall within its domain. And while this may cover quite a lot of morally desirable interpersonal transfers, most persons would regard it as not exhausting them. Your giving someone a hand with his heavy shopping-bags is unquestionably

I am thankful to Justine Burley, Jerry Cohen, Charles Erin, John Harris, Susan Hurley, and Jonathan Wolff for their comments on several of the arguments advanced here. I should also like to thank Oxford Amnesty for their invitation to participate in this lecture series, and I'm especially grateful to Justine Burley for all the trouble she's taken to arrange it. Second, I should say that this wonderful phrase, 'silver spoons and golden genes', is not one of my own devising: I owe it to Thomas Nagel who used it in a lecture he gave here a couple of years ago: Thomas Nagel, 'Justice and Nature', H. L. A. Hart Memorial Lecture, Oxford (1996).

a morally good thing to do. But in the absence of some sorts of special circumstance, we would tend to regard your failure to do so as an unkindness rather than an injustice.

Theories of distributive justice are accounts of the reasons we can plausibly and coherently offer for these getting and giving judgements. And the subject of this lecture—talent differentials and distributive justice—is motivated by the common view that at least some of what people should get or give should have at least some connection to what they do. Now what they do obviously depends, in part, on what they can do: on what their talents or abilities are. And what their abilities are in turn partly depends upon what was previously done to or for them when they were children. Hence the focus of this lecture is on how our thinking about distributive justice can be brought to bear on problems surrounding the formation of children's abilities.

Children have always posed a big problem for political philosophy in general and theories of justice in particular. Just when—at what age—a person's minority ends, will presumably remain a perennial subject of legal and political debate.[1] But we can all agree that persons below some age or other are definitely minors. As minors, they are assumed to lack enforceable duties and thus to be incapable of giving binding consent. They don't do social contracts, actual or hypothetical. And they are not the ones who are liable to legal action for whatever harm their behaviour may cause to others. Whether they are therefore better understood to lack rights altogether or, alternatively, to have rights which various sets of adults exercise on their behalf, is a matter of complex and longstanding philosophical dispute and one which I hope to side-step here.[2] That is, I hope to resist the very considerable pressure which a topic like the present one inevitably exerts for adopting a position on this undeniably important issue.

A major source of that pressure is the tendency of our moral thinking to operate within a conceptual framework which, following Kant, exhaustively divides the contents of the world into two categories: persons and things.[3] Things—rocks, plants, machines, arguably members of other species—are items which we can permissibly treat merely as means to our own ends. But persons are not. Persons are, in some elusive sense, ends in themselves. They—at least adult persons—are repositories of agency and, as such, are the generators of whatever is morally good or bad. And their being so is something that must be taken into account when we interact with them in the course of pursuing our own ends. Just exactly what this 'taking into account' amounts to is another matter of deep controversy. But it is fairly common ground amongst the controversialists that a broad description of it includes the idea that, in pursuing those ends, we mustn't violate others' just rights, either through direct force or, more indirectly, through exploitation. Regardless of how worthy our ends may be—and even, indeed, when they strongly include concern for the well-being of those others—we're not allowed paternalistically to impose these upon them. Persons must be treated as what Kant called 'their own masters' or, in a more recent phrase, as self-owners.[4] Practices like slavery and the Nazi eugenic programmes are only the most dramatic examples of not treating persons as self-owners.

Indeed I would argue that, unfamiliar as this notion of

self-ownership is (outside the confines of political philosophy), it is this right that seems to supply the underlying justification—and, especially, the demarcation—of such more familiar human rights as those against murder, assault, rape, arbitrary arrest and detention, interference with free speech, free contract, and free association, and so forth. Why does my right of free speech not give me a protected liberty to shout in your ear? Why is interference with an act of rape not a violation of the rapist's right of free association? Why is consensual medical surgery not a form of assault, whereas forcible prevention of a sane person's suicide is? Questions like these each have their own particular answers, of course. But as soon as we press on those answers and ask why the human rights they invoke do or don't count as reasons for allowing the actions involved, we're invariably led to something like each person's self-ownership as the common underlying right that both sustains and specifies them. For it is this right that is needed to reconcile and render mutually consistent the various liberties and powers afforded by those traditional human rights: rights which would otherwise encumber us with numerous incompatible demands.[5] And an important implication of persons being self-owners is that, in the absence of any wrongdoing, they must not be subjected to any involuntary servitude. So their talents and, for that matter, their body-parts must not be conscripted into the service of others.

Now this obviously places serious constraints on the range of permissible forms of redistribution that can be undertaken to overcome differentially distributed well-being. Vitally aware of glaring interpersonal inequalities, both within and between societies, we're naturally—and

I think rightly—drawn to the view that their reduction is morally imperative. We think it is desperately wrong that some people lead hungry, homeless, and disease-ridden lives, while others enjoy an abundance of consumer goods and services. But the requirement that we respect persons' self-ownership imposes substantial restrictions on how we may go about doing this. It tells us that, although all voluntary forms of such redistribution are entirely permissible, the same is not true with regard to many forms of enforced redistribution. We can and must compel those who harm others to pay them compensating damages. But we cannot permissibly compel those who are well endowed with talents or abilities to transfer goods or services, produced by those abilities, to those who are badly off. Not only is direct conscription of their abilities prohibited, but so too is the taxation of the fruits of those abilities.

Silver Spoons and Golden Genes

The core of the argument, connecting an exclusive entitlement to one's abilities to an exclusive entitlement to the products of one's abilities, is not all that hard to construct. If you devote time and effort to developing some proficiency in mathematics or a robust cardio-vascular system, you're better placed to win certain darts games or to go mountain-climbing. It seems implausible—and certainly contrary to self-ownership—that these benefits you secure from exercising those abilities should be morally liable to taxation. And so the same implausibility seems to attach itself to the taxation of monetary income streams that flow your way because, instead of deploying those abilities in darts-playing or mountain-climbing, you find it more beneficial to exercise them as a professional actuary or athlete. Exchanging the products of your abilities, rather than consuming them yourself, doesn't seem to increase their moral liability to taxation—at least, not in any obvious way. Self-owners are each responsible for their own choices and cannot justly compel one another to undo the distributive consequences of those choices. So if this argument is correct, there's not much permissible inequality-reduction to be had along these lines. Differential holdings of artefacts, the things we unequally produce in exercising our unequal abilities, do not legitimately occasion enforceable redistribution.

A more promising approach to the reduction of well-being inequalities begins by directing our attention to the fact that those unequal abilities cannot be developed from nothing. Self-owning adults who develop them must construct them on some already-present foundational ability which was laid down prior to that construction and prior to their becoming adults. And speaking very broadly, we can say that the factors entering into the production of this foundation are of two types: initial genetic endowment and an enormous variety of post-conception inputs which most obviously include gestational, nutritional, medical, and educational factors. The quality of these two types of production factor varies considerably from one minor to another, as do the levels of ability—or disability—that are producible from them. And those ability-level rankings generate corresponding valuations of the factors jointly involved in producing abilities: the initial genetic endowments and the diverse post-conception inputs. Here, I suppose, we loosely tend to follow the economist's way of thinking about production functions, and we say that one genetic endowment is more enabling—more golden—than another if the cost of the post-conception inputs needed to produce a given ability level with it is less than what would be needed with the other endowment. Correspondingly, one set of post-conception inputs is a more silver set of spoons than another if, in combination with any given genetic endowment, it produces a higher ability level than the other. So the lives of differently reared identical twins have often served as a testing-site for the quality of different kinds of spoon.

Now, the question we want to address is one which asks how a conception of justice, that takes the Kantian ends–means injunction seriously, can allow—let alone require—us to reduce interpersonal differentials in ability-level. Can we justly employ legal measures to constrain these inequalities? We've seen that justice prohibits us from enforceably reducing differentials in well-being. Can it take the opposite view with regard to inequalities of ability? I think the answer is yes. But I also think that how it can do so is significantly altered by central aspects of what this lecture series calls 'the genetic revolution'. So let's begin this part of our exploration by persisting a bit longer with our pre-revolutionary understanding of what's involved here.

One thing that seems clear is that, under both our pre- and our post-revolutionary understandings, the items we're calling post-conception inputs—spoons—all look like counting as artefacts. And you don't need to have undergone pregnancy or to have raised children to understand why. Virtually everything that goes into the

production of us, following conception, is something supplied by our parents or by people elected or employed for
that purpose. They are the ones who supply us with
our pre-natal environment, our medical care, our food, our
schooling, our music lessons, football kits, and all the rest
of it. It is they who largely determine the enabling
quality of these inputs. And increasingly these days, where
that quality is deemed to have fallen short of some acceptable standard, it is they who may be held legally responsible
for those shortfalls. Thus legal actions have already been
brought or proposed, not only against parents or guardians
for smoking or drug-taking during pregnancy, for child
abuse, and for failing to curb truancy and aggression, but
also against schools for educating badly.[6]

Now whatever we may think about the merits of such
litigation—personally, I think it is to be welcomed—what
it more generally signifies are three points of great relevance to this discussion. First, it supports or presupposes
the view that post-conception inputs are indeed artefacts:
they and their resultant impacts on minors' ability levels are
the products of particular persons' efforts and negligence.
Second, it is those producers—rather than the public in
general—who are liable for damage inflicted by their products. And third, the product standards to which those
producers are held, though certainly not sufficient to
ensure that each minor receives an equal amount of these
inputs, are none the less meant to ensure that each receives
some minimum amount of them. That is, these standards
effect to constrain differentials in the silver-ness of the
spoons delivered to different minors. It is the existence of
such standards that gives at least sense, if not always ver-
acity, to any complaint along the lines that 'You [i.e. some

relevantly placed adult] wrongly deprived me of the higher ability level I would have had on my arrival at the threshold of adulthood'.

Justice in the Pre-revolutionary World

It is a highly significant aspect of our pre-revolutionary understanding that this same complaint is precisely not one which can be made in respect of lower ability levels which are attributable to poor initial genetic endowments. I cannot say to my parents 'You wrongly deprived me of the higher ability level I would have had on my arrival at the threshold of adulthood, because you conceived me from gametes or germ-cells carrying genetic information which severely constrains my mathematical or cardio-vascular capability.' Why can't I say this to them? Not because it is not true: for all I know, it is. But even if it isn't true, that is not a reason why I cannot say it to them, but only why I should not.

No, the reason why, pre-revolutionarily speaking, I can't say this to them goes in two steps, as follows. First, it is an accepted feature of personal identity that if my parents, instead of using that pair of their germ-cells to conceive a child, had used another pair—another sperm and/or another egg—then that child would not have been me: it would have been someone else. Indeed, the different pair of cells they used four years later did produce someone else: namely, my brother. Let's call this the *Precursor Condition* of personal identity. And second, if—to be me—I had to have been conceived from that particular pair of precursor cells, then, to be me, I had to have had the particular initial genetic endowment I did have. There was no way my parents could have conceived me without thereby creating a zygote with that endowment. Let's call this the *Genetic Condition* of personal identity. I had to be a person with, say, a severely genetically constrained mathematical or cardio-vascular capability. The presence of that genetic disability in me is not the doing of my parents—it is not an artefact—but rather simply a piece of bad luck dealt to me by nature. Indeed, it is not strictly dealt to me, in the way that an impoverishing poker hand or a crippling lightning bolt is dealt to me. Rather it is actually a constitutive aspect of me: part of who I am, in a way that the effect of being hit by lightning or bad poker hands—or even, for that matter, irradiation as a fetus—would not be part of who I am.

So, however disabling my initial genetic endowment may be, there's no point in my contemplating bringing a charge of what's sometimes called 'wrongful life' against my parents: one alleging that they've harmed me because I would have been better off conceived differently. It's true that wrongful life charges have indeed been brought in some instances of severely genetically disabled children. But even in the few such cases that haven't been dismissed, the plaintiffs—the persons considered to have been wronged—are invariably the parents themselves, and the defendants are various persons who allegedly failed to provide the standard of pre-conception genetic counselling expected of them. And any damages awarded have been based on parental suffering rather than the remedial costs of the disability borne by the child. The child's genetic disability is not deemed to be anyone's fault: it isn't a tort committed against the child.[7] Rather it is essential to that child's identity and is entirely a natural precipitate. What, then, are the just distributive implications of this?

In the otherwise highly diverse range of justice theories associated with the Kantian ends–means injunction, whereby persons are their own masters, one premise that is shared by all of them—though often only implicitly—is that something's being a natural object gives it a different moral status from that of artefacts. The previously mentioned Kantian twofold division of the world's contents— into persons and things—further requires a subdivision of the 'things' category into artefacts and natural objects. Differential holdings of artefacts do not, as we've seen, warrant their egalitarian redistribution. But differential holdings of natural objects do. Unequal possession of what we can comprehensively call 'natural resources' occasion a justly enforceable duty in those who have more valuable ones proportionally to compensate those who have less valuable ones.

Where does this duty come from? How is it justified? Some have argued that it is based on something like John Locke's seventeenth-century theological premise: that God gave the earth to humanity in common, rather than only to divinely anointed monarchs.[8] Alternatively, that duty can be derived from some more secular set of principles.[9] In either case, the thought is simply that no one has any stronger prior moral claim than anyone else to the exclusive use of those things which are not artefacts. And consequently, everyone owes everyone else an equal share of the value of whatever such natural things he or she claims exclusively. (Perhaps I should add, parenthetically, that I've so far found no compelling reason for thinking that this reference to 'everyone' can be anything other than global in scope. In which case, the per capita equalization mandated by that distributive requirement would be an international one—possibly in the form of each person's equal entitlement to what's currently called an unconditional basic income.[10])

The concept of 'natural resources' itself, of course, covers quite a lot of territory—literally! At its most compendious, it includes all geographical sites as well as the spatial locations above and below them: that is, it includes all portions of the earth's surface, all subterranean natural resources, and things like air-space, the ozone layer, the electromagnetic spectrum, and so forth. Natural resources further comprehend those elements of the biosphere that are not the products of human labour including, as the Darwinian story implies, vast amounts of encoded genetic information. Most of these natural resources are used, and used extensively, by us in the course of our various activities. And many of them are bought, sold, and leased with considerable frequency. Accordingly, they clearly have an assessable value. And justice requires that value to be distributed equally.

So, telescoping a rather extended chain of reasoning, we arrive at the conclusion that—*ceteris paribus*—the egalitarian proviso on natural resource distribution requires that those who have children with golden genetic endowments owe net transfer payments to those who don't. And those who don't are thereby supplied with the wherewithal to invest in remedial spoons which are sufficiently silver to raise the lower ability levels that would otherwise result from their children's poor genetic endowments.

Since ability and disability levels are identified and measured relative to some social mean, and since the valuations of genes and spoons are determined on that basis, we might reasonably expect that, under such a redistributive arrangement, these compensation flows would considerably reduce differentials between children's ability levels at the threshold of adulthood, when the further development and deployment of their abilities become matters of their own choice.[11] Arguably, those current public policies which make special educational and medical provision for disabled children, represent some highly imperfect approximation to at least the expenditure (though not the taxation) aspects of this sort of redistributive arrangement. Perhaps such policies even signify some very fragmentary recognition of the more general idea that the costs of those adversities which are due to nature—rather than to the activities of particular persons—should be borne by all of us and not just by those whom they befall.

That, it seems to me, is how justice addresses the problem of talent differentials under our pre-revolutionary understanding. So how does our post-revolutionary understanding mess this story up? Probably in lots of ways. But I want to focus on what I take to be the central one.

The core of what it does, I think, lies in its impact on what I previously labelled as the Genetic Condition of personal identity. This condition stated that since, to be me, I had to have been conceived from the particular precursor cells I was conceived from, then—to be me—I had to have the particular initial genetic endowment I did have. My genetically constrained mathematical or cardiovascular ability is an essential constituent of my identity. So I cannot intelligibly complain that I would have been better off had I been conceived without it, and that I was therefore harmed by being conceived with it.

Justice in the Post-revolutionary World

Now, the radical move that our post-revolutionary understanding makes is precisely to sustain, rather than deny, the intelligibility of that kind of complaint. It does this by breaking what we pre-revolutionarily believed to be the necessary empirical link between the Precursor Condition of personal identity and that Genetic Condition which it simply discards. It tells us that although, to be me, I certainly and necessarily did have to be conceived from those particular cells, there was no causal necessity about them carrying the particular genetic information they did carry. They would still have been the same cells—*their* identity would have been unchanged—even if, prior to my conception, the components of that mathematically or cardio-vascularly disabling information had been snipped out of them and replaced with something less disabling. So, without falling foul of the Precursor Condition of personal identity, I can reproach my parents for harming me.[12] Or rather, such reproaches will increasingly be able to be made as our knowledge of the human genome advances and, with it, our capacity to manipulate its information loads.

What the genetic revolution does is to give a massive shove to the always moving frontier between nature and nurture: it puts events and objects which we've long treated as natural firmly into the domain of choice. Our distinction between genes and spoons gets drained of the moral relevance I previously attributed to it—genes become the moral equivalents of spoons—and our initial genetic endowments become re-classified as artefacts.

Does this threaten our capacity to reduce children's ability-level differentials, while still remaining within the parameters of what justice permits? Recall that a substantial portion of that capacity rested on our pre-revolutionary assumption that initial genetic endowments are natural objects. But if it is now, or is about to be, the case that they are artefacts, can their unequal distribution warrant any compensating redistribution? After all, unequal holdings of artefacts don't, in themselves, permissibly occasion enforceable transfers from those who have more of them to those who have less.

I think the answer to that latter question can be 'yes'. Yes, we can justly require such compensating redistribution on our post-revolutionary understanding. But the direction or form of that redistribution must be different. The genetic revolution, in forcing us to discard the Genetic Condition of personal identity, shifts responsibility from nature to particular persons. Natural genetic adversities become torts and, accordingly, public liabilities become private ones. For if genes are spoons, then the same sort of product quality standards that currently apply to the post-conception inputs entering into the production of children's ability levels, can equally apply to the pre-conception ones. Post-revolutionarily, we can say that some conceptions harm the persons thereby conceived: they place them below a minimum level of genetic advantage which those same persons could otherwise have enjoyed. Consequently, we're able to make sense of the notion of 'wrongful life' and to vest persons with rights against it. Just as they can sue schools for poor education, so too can they be empowered to sue for having been endowed with disabling genes. And presumably the damages awardable in such cases would bear some close relation to the cost of the remedial spoons needed to offset that genetic shortfall: remedial spoons which might well come to include various forms of somatic cell gene therapy.[13]

I won't pause here to explore the sorts of incentive structure that such a right might predictably bring to bear on people's procreative choices. Perhaps these are obvious enough. And as far as I can see, they all seem to press in the right direction. What may be worth some brief mention, not least because it has figured elsewhere in this lecture series and continues to attract a good deal of anxiety about the genetic revolution, is the issue of cloning. Should the foregoing account, of distributive justice with respect to genes, cause us to worry about the prospect of human cloning?

I can't really see why. None of the arguments advanced here has been particularly concerned with the kinds of precursor cell we use to conceive persons. The same liability rules apply, so to speak, regardless of whether we do that exercise by combining two whole germ-cells or, instead, by combining an enucleated germ-cell—an egg— and a somatic cell with induced totipotency and supplied by either the same donor or a different one. Any technique for conceiving persons seems, as far as justice goes, to be as permissible as any other. The standard worry about cloning is that it produces interpersonal uniformities and a consequent loss of various dimensions of human diversity. But if there are morally relevant differences—between those uniformities which are genetically induced and those

which are induced, say, by a national educational cur-
riculum—they aren't differences which concern justice.[14]
It would be difficult, to say the least, to sustain a claim
that an identical twin is, *ipso facto*, a victim of injustice.
What post-revolutionary justice does need us to worry
about is any conception that imposes genetic disadvantages
on the person thereby conceived. And there's no particular
reason to suppose that cloning would do this either more
or less than any other conception technique, ancient or
modern.

Conclusion

Let me close this chapter, then, on a slightly upbeat note
about human rights. In the recent Hollywood comedy
Mars Attack! there's a moment when that arch-portrayer
of boisterously cynical characters, Danny de Vito, remarks
'If you wanna conquer the world, you're gonna need
lawyers—right?' What de Vito is telling us here, I take it,
is that the transaction costs of such ambitious projects have
increased substantially since the days of Alexander the
Great and Attila the Hun. And I don't think it is over-
sanguine to suppose that a major factor in that increase
has been the fact that it is no longer quite as easy, as it
once was, simply to roll your elephants or your tanks or
your secret police over other people's human rights. That
set of constraints on oppressive projects has, especially in
recent years, become very much more real and far less
negligible than it used to be. And Amnesty has had a lot
to do with that.

What that set of rights is meant to protect us against
are various forms of harm and disablement at the hands

of others. So, to conclude, it doesn't seem utterly fanciful to imagine that—before too much longer—a right against genetic disablement might come to be seen as a proper member of that set.

Tin Genes and Compensation

Jonathan Wolff

One highly prominent theme in libertarian thought is that of privatization. Libertarians wish to reduce the role of the state or society to a minimum and Hillel Steiner is no exception. However, as a libertarian who is also an egalitarian, the framework in which he operates is rather distinctive.

Many egalitarians make a basic distinction between 'choice' and 'circumstance'. On this view one is respons-ible for those results of one's freely chosen actions or decisions, but should receive compensation (or pay tax) for the results of bad (good) luck. Compensation should be received from 'society' and society will receive the tax dividend from those who benefit from their good fortune. On such an egalitarian scheme those with genetic disad-vantages should receive compensation from society for this disadvantage.

According to Steiner, however, the basic egalitarian distinction between choice and circumstances is too sim-plistic, and this has important ramifications for 'post-revolutionary' genetic justice. Steiner has argued that we need a three-way distinction. Choice and circumstances should be supplemented by a further category: things that

My thanks to Hillel Steiner for comments on an earlier draft, and to Justine Burley for inviting me to make this reply.

happen to you which are caused not by your own free choices, or circumstances dealt by nature, but because of the free or negligent actions of others. This is the realm of tort. Thus Steiner agrees with the egalitarian that one should be compensated for adverse fortune if one is not responsible for it. But the libertarian in him argues that it is not true in all cases that compensation should be a general charge on society as a whole. Rather, where one's poor fortune is a result of a particular person's, or combination of persons', free action or negligence, then liability for compensation falls squarely on that person or those persons. This we can see as the project of privatizing harm or wrong. Where individuals are at fault only those individuals owe compensation. General redistribution is replaced by tort law.

The relevance of this to the question of compensation for genetic disability should be clear. Children are not responsible for their own genetic disability. Therefore they should receive compensation. Who from? In the traditional egalitarian framework the only answer is society, i.e. the taxpayer at large. But in Steiner's alternative framework we can ask whether any particular individual is at fault: if I am born with tin genes can I blame my parents and seek compensation from them?

According to Steiner, on our 'pre-revolutionary' understanding of genetics, this question apparently involves some sort of metaphysical absurdity. Can I blame my parents for harming me by allowing me to have been born with defective genes? Well, if they had not used that particular sperm and egg, with their particular genetic make-up, then the child they would have produced would not have been me. I could not have been born without my particular genetic endowment—defects and all—and therefore there is no sense in which I have been harmed by them: there is simply no standard against which I can measure to say that I have been harmed.

In the post-revolutionary world, however, there are important changes. We realize that we may be able to alter part of an egg or sperm's genetic make-up without changing the identity of the subsequent individual born from that combination. Hence the defective sequence of genes could have been repaired, and I would still have existed. Hence I can say that my parents have harmed me by not carrying out the relevant gene therapy. So any compensation due to me should be paid by my parents, whose fault it is, rather than the taxpayer. Thus the wrong of genetic disadvantage is privatized: it becomes a tort.

This is a fascinating argument, and no doubt many lines of prima facie objection are possible. Here I want to try out two; one which considers the relevant counterfactual for harming and the other which points out an internal difficulty for Steiner. Finally, I will suggest another way of arriving at a similar conclusion while avoiding some of the difficulties.

Suppose we agree that sooner or later a variety of options will be available to deal with potential genetic abnormalities. Already we have genetic counselling where couples are advised on the statistical probability that their offspring will suffer various disabilities. We also have limited genetic screening, where embryos known to be at risk can be tested, and a decision can be made as to whether to terminate the pregnancy. Two further developments—currently the stuff of science fiction—may be achieved in the future. One we could call 'germ-cell selection': the genetic material of a sperm and egg could be scanned, and a complete account of its genetic make-up determined, before it is used to produce offspring. Defective cells would be rejected and normal cells selected. Finally, there could be 'germ-cell surgery' in which a particular sperm and egg are chosen, and then one or both are genetically altered by mechanical intervention, to remove any potential genetic deficiency.

It seems to me that Steiner's argument for privatization only becomes relevant when germ-cell surgery is a real possibility. But, abstracting now from questions of pro-ducing the *Ubermensch* and concentrating only on rectifying disability, why would any prospective parent choose germ-cell surgery over germ-cell selection which would presumably be far cheaper and easier? One possib-ility, of course, is that a couple may have only defective cells, or so few non-defective ones that they are impossible to find. But in more usual circumstances there will be a mixture of defective and non-defective germ cells, and selection will be preferred to surgery.

Why is this relevant? To consider the question of whether I have been harmed by being born with defective genes we have to ask 'harmed in comparison to what?' We saw that on the pre-revolutionary view if I had not been born with defective genes I would not have been born at all, so there is no sense in which I have been harmed. On the post-revolutionary view, according to Steiner, the possibility of germ-cell surgery means I can conceive of having been born without disability, and hence we have a bench-mark for harm.

The complication, though, is this. In most, if not all, cases prospective parents would opt for germ-cell selection, rather than germ-cell surgery, if they had considered taking any genetic precautions. Thus the closest possible world in which I do not have my genetic disability is once more one in which I am not born at all. Why? Because in the closest possible world in which my parents take genetic precautions the particular sperm and/or egg, on which my identity is reliant, would have been rejected entirely, along with its genetic defect. Therefore, in such a world I simply do not exist. Once again there is no bench-mark for harm. Steiner considers the case where the defect is remedied. He does not consider the far more likely scenario in which cells found to be defective are simply passed over until a sound one is found. Thus it remains the case that in close possible worlds I do not exist without the genetic disadvantage. So, we are in the same position as we were on the pre-revolutionary under-standing. While I can blame my parents for having a child with defective genes it remains very unclear that I can blame them for allowing me to be born with defective genes.

We will see shortly why we might welcome this objec-tion to Steiner's argument. Before we do that, though, I want to raise the second, internal, difficulty for Steiner. This can be stated rather easily. Although Steiner says that he wants to remain neutral on the question of whether children lack rights altogether, or as having rights which adults exercise on their behalf, it seems clear that he needs to assert that children have at least one right: the right to sue their parents for failing to engage in germ-cell therapy.

However, in n. 2 Steiner mentions that he argues else-where that minors do not have rights, and indeed, elsewhere he argues that during their minority children are the property of their parents. The difficulty here is obvious. If children are the property of their parents then any rights they appear to have are properly vested in their parents. Therefore the parents are properly the plaintiffs in any case for compensation for their children. But unfortunately, on Steiner's scheme, they are the defendants too. So in effect parents will be suing themselves for their own shoddy workmanship: a prospect to be welcomed only by lawyers.

Therefore, unless Steiner wishes to grant minors independent legal standing they will have to wait until they reach majority to sue their parents. Possibly by this time their difficulties will be beyond remedy, or at least beyond easy remedy. For all Steiner's ingenuity his argument seems to come to little. And even if we allow children the right to sue their parents it is questionable what this would achieve: typically, parents of disabled children devote a large part of their time and resources to those children in any case. There may just be no further assets to call on, unless they were insured. But if they were thoughtless enough not to take available genetic precautions, how can we suppose they would be thoughtful enough to take out insurance?

It seems to me, though, that a more humane form of 'semi-privatization' is possible without engaging in the metaphysics of identity, without waiting for the genetic revolution to be complete, and without having to settle the question of children's rights against their parents. All we need do is to realize that we can break down issues of compensation and liability into two stages, while remaining within Steiner's three-way framework of individual choice; nature or circumstance; and private wrong.

Consider, first, the child with a genetic disability. Suppose for the reason given above we are unconvinced that that particular child can claim to have been harmed by being born with that disability: in close possible worlds, no disability, no child. However, the child suffers a disadvantage which should be compensated for. Here the only possibility within the framework is that the charge should fall on society as a whole.

Consider, now, the parents. The fact that they did not harm this particular child does not entail that they have not been negligent. Suppose they took no genetic precautions whatsoever. They are responsible, through negligence, for bringing into the world a child who requires compensation by society. Therefore, within Steiner's framework, they owe to society an exactly similar sum to that which society is paying to the child. The parent pays; the child receives; but society mediates the relation.

I called this a more 'humane' solution, and the reason is this. Under Steiner's solution, where the debt is owed directly from parent to child (ignoring my second objection) should the parent default, or be unable to pay compensation, the child, presumably, is left uncompensated. However, on the solution proposed above the compensation payable to the child will be paid irrespective of whether that child's parents can pay their debt to society. I do not claim that this scheme is the correct approach to genetic disability—although it may be—but it seems to me that it has strong advantages over Steiner's own proposal, and this is quite independent of the difficulties we have found with that proposal.

A Perspective from Africa on Human Rights and Genetic Engineering

Solomon R. Benatar

Eric Hobsbawm has described the twentieth century as 'The Age of Extremes'.[1] One extreme is exemplified by the many scientific and technological advances which have increased our life-span and improved life in a myriad of ways. Advances in molecular biology and the promises offered by genetic engineering to benefit human life typify the success achieved through the scientific reductionist approach to micro-structures. The other extreme is reflected in the magnitude of human suffering flowing from poverty, belligerence, and disregard for nature. The production of weapons of mass destruction with the potential to destroy all life, the extent of socially caused human suffering, and the escalating impact of modern lifestyles on the environment illustrate how short-term gains may in time become outweighed by greater losses through failure to consider effects at the macro level.

The African continent, a paradigm example of an underdeveloped region, offers another example of extremes in our modern world.[2] The successful and peaceful political transition in South Africa, from the abhorrent apartheid regime to a democratically elected government with a liberal Constitution and Bill of Rights, represents almost undreamed of success. However, this success is in stark contrast to the social and economic disarray in Africa. Intense poverty, widespread starvation, recurrent wars, infectious diseases—including escalating rates of HIV infection—and failure to achieve widespread democracy, are all manifestations of failure at the level of whole systems which cause untold misery and may in the long term undermine progress throughout Africa, including South Africa.

Recrudescence of tuberculosis in recent years is yet another example of extremes.[3] Scientific advances, which include (i) discovery of the microbial cause, (ii) understanding basic pathobiological processes, (iii) the development of new diagnostic tests, a vaccine, and highly effective drugs, and (iv) the design of cost-effective treatment regimens, represent the successful end of the spectrum. These successes which could have resulted in the eradication of tuberculosis (as was achieved for smallpox) are being eclipsed by the emergence of multi-drug resistant organisms through failure to implement treatment programmes effectively at the community level. The result could be that instead of its eradication, tuberculosis may become untreatable worldwide.[4]

In this chapter I should like to describe some of the grotesque disparities which mark the polarized state of our increasingly unjust world, and then discuss how exploitation and disregard for fellow humans have become central causes of these disparities. This will set the scene for appreciating some fears from Africa of potential adverse implications of advances in biotechnology. I shall then discuss how deficiencies in our contemporary approach to human rights may play a role in sustaining the divide between extremes or, at the least, failing to narrow them. I shall suggest that acknowledging the powerful forces which sustain social injustice is the first step towards new ways of thinking that could promote a broader notion of moral behaviour and a socially responsible concept of rights. Finally some suggestions will be made for enhancing the achievement of human rights more widely.

Extremes and Disparities

One billion people live well, some in great luxury in a rich core, and 3 billion live in profound poverty on less than $2 per day in the poor periphery. The gap between the richest and poorest quintiles of people in the world is widening inexorably and increased from 30-fold to 60-fold between 1960 and 1990.[5] In 1994, 45 per cent of the world's population (2.5 billion people) lived on 4 per cent of global GNP, while 358 billionaires owned the same proportion of global resources.

Associated extremes of biological suffering are reflected in the spectrum of disease patterns, and wide range of life expectancies, infant mortality rates, nutritional status, and living conditions around the globe. Wide disparities are also evident in the social health and suffering generated through socially determined life patterns and hierarchies of power. Witnessing the squalid conditions of life and facilities in public hospitals in central Africa evokes the feeling of being transported back into the middle ages and vividly portrays the extent to which spectacular developments in the modern era have not even remotely benefited billions of people.[6]

Disregard for Others and Exploitation as the Major Cause of
Disparities

South Africa's history and the extremes in wealth, in health
status, and in the quality of life associated with 40 years
of apartheid are well known.[7] That 20 per cent of the
population owned 80 per cent of the country's resources,
and that this was brought about by racial discrimination
and exploitation, correctly horrified the civilized world.

Human life has long been characterized by discriminatory attitudes and exploitative processes, many centuries
of which have contributed to constructing the lifestyles of
the privileged and poor worldwide.[8] With the passage
of time and the growing complexity of life, previously
overt and crude discriminatory and exploitative processes
in the battle for resources have become more covert and
sophisticated. Under such new guises national and international economic policies driven by wealthy
industrialized countries have profoundly transformed the
global economy in the second half of the twentieth
century. The minimum prices paid for raw materials
associated with devaluation of the currency of poor countries, payment of demeaning wages to workers in foreign
countries less committed to human rights, protectionist
trade practices, and the trillion dollar a day market in
foreign exchange across financial networks (only 10 per
cent of which is for trade in goods and services) have
resulted in extremes of wealth and poverty.[9] While material
conditions of life have generally improved, not all have
benefited, as wealth and misery are generated simul-
taneously.[10]

The external debt of developing countries is another
major force sustaining poverty, obstructing the develop-

ment of human potential, and creating an instability that threatens all nations. Such debt has grown by $100 billion annually, doubling from $650 billion in 1980 to $1.3 trillion in 1990. At $2.2 trillion by 1997 it amounted to 45 per cent of the gross national product of developing countries. This represents a debt of almost $400 for each person in the Third World—much more than a year's earnings for many. The ways in which such debts have accumulated (despite many noble intentions) under the influence of

(i) excessive spending by some wealthy nations,
(ii) the backlash from oil producers,
(iii) the sale of weapons to developing countries and
(iv) collusion with corrupt leaders,

are now well documented but not widely appreciated.[11]

Sophisticated forms of fraud on a large scale pervade economic activities. Insider trading, corporate takeovers, corporate greed, false information in investment promotion, and episodes of what J. K. Galbraith calls 'speculative insanity', all unfavourably affect income distribution.[12] Wealthy banks have held on to illegally acquired resources, and accepted into secret accounts deposits from corrupt despots, knowing that these resources have been diverted from the purposes intended by benefactors and that debt repayment would cripple the citizens of indebted countries.[13] The implications for countries unable to repay their debts, and the boomerang implications of the debt crisis on rich nations, give us good cause for reflection on economic policies and market forces designed to sustain unrealistic economic growth for rich countries at the expense of world ecology and some of the most vulnerable people in the world.[14]

In post-industrial countries where human services have become a major component of the economy, exploitation is extending into services that are essential for the basic well-being of citizens. In the USA, where taxation is lower than in any other industrialized country (tax revenue accounts for 32 per cent of the overall economy in the USA as compared with up to 58 per cent in some Nordic countries and up to 50 per cent in Western Europe as a whole) and the cost of petrol is one fifth of that in Europe, many services supportive of human development are being withdrawn from its most vulnerable citizens. Indeed erosion of welfare programmes is becoming the norm within a globalizing economic paradigm.[15] For example, medical care is being transformed into a marketable commodity available to those who can pay, and rationed for those who cannot. Resulting adverse social effects include the creation of widely disparate health statuses even within rich countries.[16] In the USA, reform towards universal access to health care, offered by all other advanced societies, has been thwarted by the power of those with entrepreneurial interests in health-care delivery.[17] Against such forces even authoritative advocacy for health care reform by such a prestigious organization as the American Association for the Advancement of Science has been ineffective.[18]

Other aspects of health care are also being transformed by the profit motive. International pharmaceutical companies, unsatiated by their profits in the industrialized world, are selling drugs through intermediaries in poor countries at prices higher than in the wealthy world.[19]

Counterfeit drugs are pervading the market, technology is being used excessively and inappropriately, profit-making by entrepreneurs has become a major aspect of health-care insurance, research is increasingly being driven by fiscal considerations, and doctors are being swayed away from patient advocacy by financial and other conflicts of interest.[20] Health-care systems throughout the world are fracturing under the influence of these forces which are relegating the caring and professional aspects of health care to second place. The poor example being set by some wealthy countries is having devastating effects on attempts to provide health-care services on a more equitable basis in poor countries.[21]

Economic considerations also influence attitudes to war and the industries which sustain conflict. Following the Second World War the superpowers perpetrated their ideological differences through wars in distant countries, often in collusion with despots and at the cost of many lives, while profiting through proliferating arms industries creating weapons of mass destruction and a nuclear stockpile that will pose hazards for thousands of years.[22] The extent to which the military has become autonomous in many countries (including the USA) and immune to democratic control has been described with foreboding by J. K. Galbraith. Virtually all wars since 1945 have been fought in Third World countries, resulting in 23 million deaths. More recently, civilian wars, often linked to historic ethnic enmities, opposition to oppressive governments, or arising from artificial geographical boundaries created by colonial powers, have resulted in millions of deaths. UNICEF estimates that in 1994, 30 million children died due to wars and poverty—300 times as many deaths as resulted from the bombing of Hiroshima and Nagasaki. It would seem that within a few hundred years 'enlightenment' has been eclipsed by conflict that has destroyed, maimed, and displaced millions of people.[23]

The deliberate destabilization of Mozambique by South Africa and the USA in the 1980s is an example of a war that resulted in over a million deaths and created one of the most indebted countries in the world.[24] Such ideological wars, and the structural adjustment programmes imposed by powerful financial institutions and nations have resulted in retrogression from the initial gratifying gains made in educational, health, and social services following independence from colonial powers in the 1960s.[25]

Against this background of war and economic extremes torture has become a growing problem and medical professionals are participating willingly or through coercion in such barbarous practices.[26] The torture and abuse of Steve Biko and others in South Africa has received widespread publicity, but less well known is the extent to which torture and cruelty persist in prisons in many countries. In chillingly describing cruel prison conditions in Europe, Antonio Cassese, a distinguished professor of international law, makes a forceful plea for efforts to ensure that the principles of the Enlightenment penetrate into these 'dirty worn out parts of the social fabric' in Europe.[27]

Within a globalizing economy, economic power has been shifted from potentially accountable governments within nations, to multinational corporations and other unaccountable transnational market forces. The fear of big government has resulted in the creation of big business wielding enormous power that cuts across nations. The result is a reduction of the power of nations over their own economies, the marginalization of a growing proportion of people in all countries, the defeat of liberal ideals, and the potential for implosion.[28] We now face a situation in which

(i) the combined assets of the top 300 firms in the world constitute roughly 25 per cent of the world's productive assets,

(ii) 20 per cent of the US population owns 80 per cent of the nation's wealth, and

(iii) 20 per cent of the world's population owns 80 per cent of global wealth—and the divide continues to widen.

It is interesting and disappointing to reflect on the fact that these disparities, persistence of the underlying mechanisms sustaining and aggravating them, and the threats they pose to the freedoms of individuals and the well-being of society, do not arouse intense international indignation and concern. It seems as though when others are dehumanized and exploited by sophisticated methods, covertly or at a distance, this is accepted as an inevitable aspect of life not to be resisted! Against this background we should perhaps not be surprised that Isaiah Berlin described the twentieth century as the worst century in Western history.

Some Considerations from Africa

Africa has been most adversely affected by the complex interaction between all the above aspects of modern life. It contains 33 of the 50 poorest countries in the world, and two-thirds of Africans live in absolute poverty. A brief reminder of Africa's history may facilitate empathy for its plight and the concerns of its people regarding human rights and the implications of the genetic revolution for their future.

Africa has long been central to the aspirations of the West.[29] Initially presenting great challenges to circumnavigation, it became a continent inviting exploration which resulted in a clash of civilizations, colonization, brutal enslavement of its people, and relentless extraction of its rich resources. Since the release of Africa in the 1960s to independence, powerful nations have colluded with selfish leaders co-opted into neo-liberal economic policies, and enslavement has continued to obstruct development under the more covert and sophisticated guises I have described.[30] It has been argued that in time it will be recognized that the impact of these processes, and in particular the creation of an unpayable debt, will greatly exceed the horrors of slavery.[31]

Africans must, of course, take some responsibility for the state of their continent since post-colonial independence. Poor leadership, corruption, internal exploitation, nepotism, tribalism, authoritarianism, military rule, and overpopulation through patriarchal attitudes and disempowerment of women have all contributed to its sad state, although, to be fair, these shortcomings must be seen in the context of powerful external disruptive forces acting over several centuries.[32] The tragedy of Africa is now being aggravated by its elimination from the foreign policy agendas of powerful countries and its marginalization because of crises conveniently perceived to be of its own making, and 'of such diffuseness and magnitude that the world at large shrinks from engaging them'.[33]

While for privileged people it may seem that the balance in the use of power flowing from scientific knowledge and technological achievements has been in favour of beneficence, different perceptions prevail among those who have been marginalized. Close links between science, technology, the military (defence), money, and those with global power,[34] and the use of power and secrecy to protect privilege has undermined confidence that there is any significant concern for the future of the people of Africa.

The viewpoint I shall express from an African perspective regarding modern biotechnology should be seen against this background, but not as an attempt to provide an all embracing view. Africa is a large continent, its many cultures are not static, there is some optimism for its future and it would be inappropriate to make a stereotypic case. Africa's problems have been so enormous that debate on genetic engineering has not had much of a profile there, and I can only provide a synoptic perspective that may resonate with views held by many in Africa.

When biotechnology is used to mass-produce drugs such as insulin and vaccines, to develop more resilient crops, to increase the efficiency of food production, or in other ways that improve the lives of individuals, this is uniformly welcomed, especially if costs are reduced and access increased for all to drugs, vaccines, and food.[35] However, it is possible, perhaps even likely, that such advances may not be available to those in poor countries because they are too costly—as for example with new drugs for treating HIV infection.

Worse scenarios can be envisaged. For example, techniques that are paving the way to controlling farming and world food production by giant agrochemical companies may threaten subsistence farmers. Reduction of the genetic diversity of agricultural products or excessive protection of intellectual property rights locking farmers into corporate control of production could have profoundly adverse impacts on the economies of developing countries and the lives of their citizens. Attempts to patent products developed with information derived from the practices of indigenous healers, and eagerness to patent components of the human genome for exclusive economic gain are seen as new forms of exploitation that exclude considerations of the humanity of those living in economic misery.

Recent developments in biotechnology are also contributing to the development of biological weapons. Such weapons, together with anti-personnel acoustic[36] and blinding laser weapons,[37] are adding new dimensions of military horror to future conflicts. There are also fears, as unrealistic as these may seem to some, that biotechnology will allow the development of genocidal tools to propagate racism and supremacism through eugenics in a modern guise.[38]

In describing the implications of the Human Genome Project (HGP) in the USA Nickens reminds us of the numerous historical examples of what has happened when genetics are drawn into arguments over the causes of social inequalities. He refers to the Tuskegee syphilis experiment, genetic theories of intelligence, sickle-cell anaemia, and the potential for discrimination in access to health insurance, health care, jobs, and public assistance programmes. He also reminds us that the HGP will introduce information into a 'United States social context which: (i) already has a long and disturbing history of drawing sharp distinctions among our citizens on the basis of race and ethnicity, and (ii) has a long tradition of belief in biological determinism'. He expresses concern that scientific theories and data have often been used to buttress prevailing biases, and points out that some have asked whether the HGP may create a new 'biologic underclass'.[39] These concerns extend beyond the USA into many other societies.

Secular western attitudes to the concept of the self as highly individualistic, and unconnected to the community or the spiritual world, further undermine the confidence of Africans who view people as both uniquely individual and as intimately connected by relationships to others in the present, past, and future. This African conception of humanity, which places great value on both the individual and on the collectivity of people, views the life-forces of an individual both during life and after death as incompatible with both organ donation and the cloning of an identical body. While this view, expressed through abhorrence for organ transplantation and a fear of the implications of twins for the individual soul, may be waning as African cultures transform under the influence of other cultures, the concept remains powerful. We should not be surprised or derogatory about this unless we are consistent in applying our attitudes to the religious beliefs, myths, and superstition which persist in a scientific and secular era in the west. Given the pervasiveness of a deeply spiritual and communal world-view in Africa, and the fact that many Africans perceive most sources of modern power to be used against them (or not for their benefit) by those lacking spiritual awe for life, and by those whose political and economic decisions are focused on individualism and materialism, it is not difficult to imagine why Africans should lack confidence that the ability to alter genetic structures will suddenly be used for the benefit of all humans—including them.

In summary, it can be said that, in the context of

(i) exponential population growth;
(ii) epidemics of diseases (such as HIV)—which some blame on African sources;
(iii) ongoing worldwide supremacist attitudes; and
(iv) the exclusion of the African continent from the foreign policies of powerful nations after many years of exploitation,

we should have some empathy for suspicions of Africans that research into the human genome may lead to new forms of control over groups of people who are seen to be too numerous and irrelevant to market economies.

Human Rights

The concept of Human Rights has developed from endeavours to achieve respect for the freedom, dignity, and equality of individuals. It is a vitally important concept for protecting individuals from oppressive authoritarianism and for allowing them to develop their full potential for fulfilled lives.[40] Since the Enlightenment demands for respecting human rights have succeeded in progressively emancipating many people from the control and indignities suffered under oppressive rulers, religious institutions, tyrannical governments, and private oligarchical power such as that wielded by ruthless industrialists and slave-owners in previous eras. The Universal Declaration of Human Rights and the development of International Law in the twentieth century were expected to extend the benefits of such freedom throughout the world and to enhance the quality of all human lives. While the idea of human rights is firmly established as universal, it is not uniformly interpreted. For some the concept is limited to civil and political rights. For others, the liberal ideal and human rights include the economic, social, and cultural rights considered to be essential for human flourishing in social democracies. Growing appreciation of the need for solidarity in the face of now well-known ecological and demographic dangers has extended the claim for rights to include rights to a clean and safe environment for present and future generations, and to relations between nations which could foster the develop-ment required for peace and social justice.[41] Regrettably insufficient progress has been made and we seem to be receding from, rather than advancing towards, a more just world.

We cannot be complacent about this. J. K. Galbraith reminds us that deriving comfort from high-minded debate about human rights in the contentment of privilege reflects our co-option into false images of beneficence and global progress.[42] We seem generally unaware of how high ideals have been distorted by the powerful political and economic forces that propagate injustice, and of the psychological processes that allow us to believe in a just world by detaching ourselves from the profound misery of so many.[43] I should like to suggest that while this sad state has complex foundations, to which I have been able to allude only briefly, it has also been contributed to (or at least not ameliorated) by several deficiencies in the approach to human rights.

A Narrow Concept of Human Rights

In recent years the long-standing dilemma regarding the balance of attention to the good of individuals and to that of society has re-emerged in many forms, which include an intense debate between liberals and communitarians. At the core of liberalism as an ideology lies a commitment to the freedom of individuals to be self-determining and self-governing with as little external interference as required to ensure equal liberties for all.[44] The interpretation of how this can best be achieved spans a wide range. At one end of the spectrum are those concerned about maximum freedom for individuals and markets. At the other end are those who acknowledge the need for a social contract within social democracies to ameliorate the potentially profound adverse effects of selfish individualism and unfettered material progress. While the challenges from communitarians are diverse and the complexity of the debate cannot be reviewed here, I believe that there is an analogy with the criticisms I shall make about a narrow concept of human rights. It is also relevant that in both cases solutions can be found within the framework of liberalism.[45]

When the focus on the concept of human rights is limited to the civil and political rights which are the minimal requirements of the political philosophy of liberalism it is not widely appreciated how this narrow interpretation of liberal philosophy has combined with another major tenet of liberalism—*laissez-faire* expansion of free trade—to facilitate exploitative processes[46] that obstruct the widespread achievement of human rights by impairing the ability of many to acquire even the most basic resources for living minimally decent human lives.[47] The triumph of freedom in economic life, in the thrust towards a global economy, has created a huge zone of unaccountability that results in defection from other liberal values, such as those supportive of community and social reform. The concern has been expressed that this may foster the unfolding of a tyranny of unaccountable economic rulers—that is to say economic totalitarianism in place of the political totalitarianism that was feared by Hayek as the road to serfdom.[48] This increasing dominance of an economic way of thinking is illustrated by comments from some US economists that 'Two thirds of the world's population are . . . superfluous from the standpoint of the market. By and large we don't need what they have; they can't buy what we sell'[49] . . . and 'In the perspective of world capitalism as we know it these people just do not count . . . Unless that part of the world develops the capacity for terrorist blackmail they will be in the charity ward for a long time.'[50] Such considerations remind us of the danger to the world from violence among the poor and dispossessed and the potential for mass migration and social chaos when poverty and disaster cause law and order to disintegrate.

While liberalism may be the only comprehensive and hopeful vision of world affairs, it is not of necessity only tightly linked to capitalism. Some who are committed to the ideals of capitalism recognize that the perception of this tight link has been formed in a period that is now moving to a close, and that there is a need to search for new ways of understanding that go beyond believing that behaviour driven by economic considerations alone can be the order-generating force for any society.[51]

From the human rights perspective it is necessary to appreciate that one of the fundamental underpinnings of the international consensus on human rights norms is that the two major categories of rights—civil and political rights on the one hand and economic, social, and cultural rights on the other—are interrelated and indivisible.[52]

Neglect of the Conceptual Logic of Rights

In popularizing the concept of human rights there has also been a failure to emphasize that the conceptual logic of rights includes reciprocal duties. Unless duties are honoured, claims for rights cannot be actualized in practice. Who will meet the co-relative duties if members of society see themselves only as holders of rights? Surely all members of a community must be seen as holders of rights and bearers of duties? As a duty-bearer each citizen enjoying rights is responsible for respecting and protecting the rights of others and for promoting conditions conducive to the implementation of human rights. This applies within nations and between nations. Human rights cannot be satisfied in poor countries if rich countries do not honour their duty not to exert economic and military power in ways that undermine the rights of others.

Inconsistencies in Monitoring Human Rights Abuses and in Applying Human Rights

Since the mid-1960s several western democracies have monitored and assessed the human rights performance of other countries. The United States is notable for two aspects of its programme. First, it selectively monitors and reports on human rights in other countries but fails to monitor/report human rights abuses within its own borders. In so doing there is denial and masking of the extent to which the freedoms of its own citizens to control their lives have been eroded by powerful economic forces. This allows human rights assessments to become vulnerable to ideological manipulation by a country wishing to promote its own political system while criticizing others.[53] Second, the exploitation and control of the economy of poor countries, which obstructs the development necessary to provide conditions conducive to respecting civil and political rights, is further aggravated by setting human rights standards for itself that exclude socio-economic rights. By ignoring the impact of its economic and foreign policies on the rights of citizens of other countries the USA seems oblivious to the damage done to the concept and the achievement of human rights by its own actions.[54] The propagation of human rights without respecting these in the universal manner intended, and assessment of human rights standards inconsistently by a country which most actively espouses the concept, also fails to instil confidence in those who have long been oppressed, and leads them to fear that the economic and security interests of the rich will always take precedence over humanitarian considerations.

This fear extends to the potential use of genetic engineering, notwithstanding UNESCO's Universal Declaration on the Human Genome and Human Rights, and the statement by the United Nations Convention on Biological Diversity that

the recognition of genetic diversity of humanity must not give rise to any interpretation of a social or political nature which could call into question the inherent dignity and . . . the equal and inalienable rights of all members of the human family, in accordance with the Preamble to the Universal Declaration of Human Rights.

The Rights of Future Generations

The paucity of concern for the human rights of future generations is apparent in the face of the ecological degradation which proceeds unchecked as citizens of the industrialized world generate ten times as much (or more) waste per capita than the poor, while calling for measures to protect the environment in the developing world. For example, the USA contains 5 per cent of the world's population but consumes 24 per cent of global resources, while China contains 20 per cent of the world's population and consumes 10 per cent of its resources.[55] Some industrialized countries acknowledge their excessive consumption and urge environmental and population control but show scant concern for the needs of others by focusing on control of population and pollution of the environment in poor countries, without making a serious commitment to reducing their own unsustainable patterns of consumption.

Garrett Hardin, the evolutionary biologist, wrote in 1968 of two types of rationality: one which strives for the good of the individual, and one which strives for the good of the group. He acknowledges that these are in inevitable conflict, and warns that when individuals are unaware of their joint plight, and blindly go about their business as though they lived in isolation, their

defections from social responsibility are re-created many times over on a grand scale . . . These individual decisions about the futility of working actively towards the good of humanity

178

amount to a giant trend of apathy, and this multiplied apathy translates into insanity at the group level. Narrow minded selfishness will drive us to extinction unless we band together and form strong organisations dedicated to global ecological goals.[56]

Continuing advances at the micro level, at which science is directed, are clearly necessary and desirable for the great benefits they offer. Indeed scientific progress may offer as yet unimagined future benefits. Similarly, endeavours to protect the rights of individuals must be sustained. However, it is being more widely appreciated that whole communities also need protection and that global instability is the price paid for ignoring long-term effects at the macro level. Little has been done since the 1992 Earth Summit to implement Agenda 21 and the costs of inaction grow. Unrelenting environmental abuse is thus leading to profound changes within our ecosystem which facilitate the recrudescence of infectious diseases such as malaria and cholera, the emergence of many new infectious diseases, relentless land degradation, rising sea levels, changing temperature gradients, and shifts in the ecological balance which will affect food production. Failure to consider the ecological costs of industrial and personal consumption patterns constitutes a 'tragedy of the commons' on a grand scale.[57] These threats to future generations will not be confined to poor countries (although they will suffer earlier and more intensively) and the spread of infectious diseases will be enhanced by travel, emigration, and refugeeism.

Some Directions for Change

Is it possible to change the unstable state of a world
threatened by polarizing extremes? It would seem to me
that at least some progress could be made, and is even
now being made, in this direction. Despite the many
failures to foster international peace, hope must be retained
for the inspired leadership that could foster wider
appreciation of the interdependence of all at a global
level, and the realization that neither rich nor poor
countries can continue to ignore internal and external
threats to prospects for widespread and sustainable
well-being.

The example set by South Africa through its dramatic
political transition supports the view that profound change
can be achieved through peaceful means. However, it must
be realized that South Africa now has to work towards
the more difficult social and economic transitions required
to narrow some of the disparities for which it was despised.
Its ability to do so will depend, at least in part, on whether
in acclaiming South Africa's success other countries have
come to learn anything about themselves and the exploitative processes that sustain their privileged economic lives.
South Africa's endeavours to diminish the legacy of
exploitation are most likely to be successful if other countries work to achieve the same ends for their own citizens
and in a global context. For example, unless the industrialized world alleviates Africa's $200 billion external debt
burden, and enables the benefits of free markets to be
matched by improvements in human rights, South Africa's
example in achieving a political transition will have much
less social and economic impact on sub-Saharan Africa

than might be hoped for. Africa may remain a continent of despair rather than becoming one of hope.[58]

Visionary insights from those thinkers who can see the world in all its complexities,[59] recognition of the crisis of liberal internationalism,[60] and, since the end of the Cold War, a better understanding of the failures of capitalism,[61] reveal how (altruism and reparations aside), rational self-interest alone should be enough to drive global progress in new directions. The time has come for a new mind-set, one in which a highly individualistic concept of human rights based on the notion of individuals as merely selfishly autonomous, is broadened into a socially respons-ible concept of human rights which recognizes both our individuality and the mutually advantageous relationships between individuals and nations.

Imaginative and ambitious national agendas and global development strategies are needed. These could be coupled with progressive deflection of resources away from military might to moral right. Diverting a small fraction of the almost $1 trillion annual military expenditure to human development could prevent more than 95 per cent of the premature deaths of over 30,000 children in the world every day in infancy, childhood, and adolescence from starvation and preventable infectious diseases.[62] Greater commitment to more equitable economic policies could lead towards living conditions that reduce popu-lation growth rates, and billions living in inescapable poverty with their energy sapped may, through hope for the opportunity to live decent lives, have their energies released to achieve this goal.[63]

Some specific and complex economic and political

activities will have to be initiated. To begin with such an approach would encompass commitments to

  (i) the recent United Nations Special Initiative on Africa;

 (ii) the World Bank's changing agenda on debt and poverty;

(iii) the Highly Indebted Poor Countries initiative;

(iv) serious endeavours towards achieving modest reductions in military expenditure;

 (v) the development of new attitudes to international relations;

(vi) alterations in consumption patterns;

(vii) implementing some form of taxation on electronic financial transactions;

(viii) and redirecting some of the resources generated in these ways towards enhancing social justice globally.

It is sobering to reflect on the fact that the cost of eradicating poverty has been estimated at about only 1 per cent of global income. Effective debt relief for the 20 poorest countries is even cheaper at $5.5 billion—the cost of building Euro Disney![64] The resources allocated to the Human Genome Project, more than $3 billion over 15 years, is illustrative of the power of the will to achieve specific goals. It does not seem inconceivable that the economic, political, and symbolic power of industrial nations could be harnessed to a moral commitment to promoting global social progress.[65]

For those who continue to believe that the 'free hand of the market' is the best way to achieve optimum human living conditions, Heilbroner provides several cogent reasons why the market system does not operate in this idyllic way. First, the units of operation within the market are no longer small, adaptable enterprises. The growth of powerful transnational companies has destroyed the stability and adaptability associated with smaller enterprises, and has created 'giant beams in the superstructure of capitalism' that destabilize national economies. Second, ignoring the cost of external effects such as pollution allows for inaccurate and irrational cost–benefit calculations with serious adverse long-term implications. Third, promotion of a culture of selfishness with emphasis on the consumption of saleable goods diminishes investment in those public goods—such as education, health care, and the environment—that are essential for the long-term survival of the social system within which the market operates.[66] Capitalism as an economic system is closely linked to liberalism as a political order and the integral connections between the two are often overlooked. However, the market, as the conduit for the energies created by capitalism, is acknowledged as imperfect and potentially subversive of some of liberalism's broader ideals. The design and implementation of new forms of control may make it less dysfunctional.[67]

It is surely clear at the end of the twentieth century that neither the concept of the individual as atomized, entirely free, and self-sufficient, nor that of the state, market, community, or culture as Leviathan can serve to enhance the achievement of human well-being in the broadest existential and physical senses. The interdependent and mutually advantageous links between individuals and nations need to be both appreciated and constructively advanced to foster a concept of human rights that is truly universal both in theory and in practice.

Work towards promoting sustainable development and renewing confidence in solidarity in an interdependent world will require a much broader approach to ethics and the moral life than can be constructed merely through a narrow conception of human rights. Human rights cannot be widely achieved while exploitation of people and nature continue unchecked. New approaches will need to go beyond the ethics at the micro level of individual relationships, to include ethical considerations at the meso level of whole populations of people within nations, and to ethical considerations at the macro level of the environment and the interests of future generations. In the quest for greater social justice it should be acknowledged that in the face of limited global resources the redistribution which all have agreed must take place in South Africa, is also necessary, and for very similar reasons, within other nations and at a global level. Galbraith's account of the good society and how this could be achieved within the constraints of our abilities is inspiring.[68] It seems reasonable to suggest that Rawls's description of the minimum requirements for a well-ordered society[69] can be extrapolated to considerations of the minimum requirements for a well-ordered world.

There is some cause for optimism that a gradual shift in mind-set is taking place towards such a new paradigm for the world. Many scholars are thinking and writing imaginatively about such issues. This is reflected in the Declaration of Human Duties, which was drafted because:

Whereas the Declaration of Human Rights represents one of the great advances of the twentieth century, it fails to address Human Duties and Responsibilities as necessary counterparts of these Rights . . . Recognition of and respect for human rights demand the acceptance of specific duties, in order to assure an adequate quality of life for all people and the persistence of a favourable environment for future generations. We therefore consider a binding responsibility for ourselves, not only as human beings, but especially as scientists and educators, to carefully and explicitly carry out these duties to the best of our ability, even if their enactment may run counter to established policies generated by traditional sources of power and influence.[70]

See Appendix 1.

Other trends in scholarly work also reveal slowly emerging new paradigms of thinking that provide hope for such progress. These changes include a deeper under-standing of history and of the social forces which shape the world, gradual shifts towards new ways of economic thinking, new directions in political philosophy, broad-ening conceptions of the self, enhanced ecological sensitivity, shifts in understanding how power can be used more constructively, greater respect for other cultures, and new perspectives on international relations.[71] It is in such changing conceptions that there is hope for wisdom in the use of new forms of power that will flow from genetic engineering and other scientific advances. (See Appendix 2.)

Future generations will judge our commitment to human rights by the extent to which we can make progress in these directions and not by how much we talk and write about human rights. My approach will be considered too idealistic by those who are satisfied with the state of the world, or who believe that we do not have the capacity to make the changes suggested. However, we should retain ambitious aspirations for human progress as all change has its origins in ideas, and history demonstrates that even spectacular change is possible. Our idealism cannot rest at the level of drawing up International Declarations, or focusing merely on abuses of individual human rights—as important as these are. We need to conceive of systems approaches within which patterns of behaviour at individual, national, and international levels could contribute to achieving our high ideals. Such ideals may be attainable if we could be more self-aware, more honest about ourselves and about our dependence on others for our existence—and more concerned about future generations.

While I am fully cognizant of the great complexity of the tasks that lie ahead I should like to (again[72]) echo the words of Chinua Achebe, the distinguished Nigerian author:

Despair should not eclipse hope . . . neither history nor legend encourages us to believe that a man who sits on his fellow will some day climb down on the basis of sounds reaching him from below. And yet we must consider how so much more dangerous our already very perilous world would become if the oppressed everywhere should despair altogether of invoking reason and humanity to arbitrate their cause.[73]

Appendix I. Declaration of Human Duties

IT IS THE DUTY OF EVERY HUMAN BEING TO:

1. Respect human dignity as well as ethnic, cultural and religious diversity.
2. Work against racial injustice and all discrimination of women and the abuse and exploitation of children.

3. Work for the improvement in the quality of life of aged and disabled persons.

4. Respect human life and condemn sale of human beings or parts of the living body.

5. Support efforts to improve the life of people suffering from hunger, misery, disease or unemployment.

6. Promote effective voluntary family planning in order to regulate world population growth.

7. Support actions for an equitable distribution of world resources.

8. Avoid energy waste and work for reduction of the use of fossil fuels. Promote the use of inexhaustible energy sources, representing a minimum of environmental and health risks.

9. Protect nature from pollution and abuse, promote conservation of natural resources and the restoration of degraded environments.

10. Respect and preserve the genetic diversity of living organisms and promote constant scrutiny of the application of genetic technologies.

11. Promote improvement of urban and rural regions and support endeavours to eliminate the causes of environmental destruction and impoverishment which can lead to massive migrations of people and overpopulation in urban areas.

12. Work for maintenance of world peace, condemn war, terrorism and all other hostile activities by calling for decreased military spending in all countries and restriction of the proliferation and dissemination of arms, in particular, weapons of mass destruction.

Appendix 2. Emerging Changes in Paradigms of Thought

Understanding History

An 'integrated systems approach'.

Shifting Economic Paradigms

From '*laissez-faire* or authoritarianism' to 'globalization' to 'freedom with accountability'.

New Directions in Political Philosophy

From 'liberalism/capitalism and communism/socialism' to 'social democracies and neo-liberalism' to 'communitarianism'. From 'rights' to 'rights and responsibilities'.

Changing Conceptions of the Self

From the 'anomic self' to the 'embedded self'.

Changing Relationships of Science, Technology, and Philosophy to Nature

From 'dominion over' to 'stewardship'.

Shifts in Power Relations

From 'knowledge and power over' to 'wisdom and power with'. From 'military might' (might is right) to 'moral example' (right is might).

Cultural Relations in an Interdependent World

From 'imperialism' to 'multiculturalism'.

International Relations in an Interdependent World

From 'realism' (power and force) to 'pluralism (order) and solidarity (justice)'.

Peace Movements

From 'acceptability of weapons of mass destruction' to their 'illegality'.

Rights and Beyond

Roger Crisp

No one could dissent from the humane tenor of Professor Benatar's marshalling of facts and arguments in his Amnesty Lecture. His theoretically impartial but practically committed approach will strike a chord with anyone who shares his horror at the atrocities committed by human beings against one another, against non-humans, and against the planet itself.

In particular, I applaud his pragmatic attitude to the language of rights: 'The time has come for a new mind-set', he suggests,

one in which a highly individualistic concept of human rights based on the notion of individuals as merely selfishly autonomous, is broadened into a socially responsible concept of human rights which recognizes both our individuality and the mutually advantageous relationships between individuals and nations (p. 181).

Benatar appears refreshingly aware of the fact that, in any construction of moral or political theory, the notion of rights cannot occupy the ground floor, at least not alone. Rights are to certain good things, or to the absence of certain bad things, and it is these goods and bads that do the normative work in justifying the language of rights. And, of course, the value of that language continues to be proved daily by the work of Amnesty International, an organization which achieves many valuable goals by bringing to bear the political rhetoric of rights against dictators and torturers the world over. Benatar maintains that what is required for the future is a broader conception of human rights, including economic, social, and cultural rights, as well as civil and political rights. He makes a strong case, but there may be some problems nevertheless.

First, the quality of economic, social, and cultural life in any society is closely bound up with the degree to which its members have a say in politics, that is, the degree to which democratic institutions and practices are entrenched within that society. Given that there is still much to be done in the political realm, it could be argued that it is too early to begin extending the notion of rights.

Secondly, many of the civil rights to which Amnesty appeals are so-called 'negative' rights, and identifying those who have the duty to respect these rights is straightforward: all of us do, including the torturers. The duties correlative to positive rights—rights to a certain standard of living, for instance, or to a particular cultural milieu—are much harder to assign, and there may be some doubt about whether creating an accepted conceptual logic for such rights would be justified. It may be better just to concentrate on the goods and bads at stake, and the individual and group-based routes to achieving the best outcome.

Thirdly, and relatedly, there is a general problem with the proliferation of rights. There may be advantages in restricting the language of rights, as did J. S. Mill, to that area of social life in which protection is required for fundamental human interests, such as freedom from severe suffering or debilitating injury. Speaking of rights to less central goods may make rights-talk less effective, and ultimately indeed pointless: one might as well go straight to discussing the goods and bads, bypassing the terminology of rights altogether.

There is also here a connected concern about conflict. One way to employ the notion of rights is to argue that we have rights to the protection of certain fundamental interests which are such that the violation of these rights could never be justified by the furtherance, to whatever degree, of less fundamental interests. This resolves many potential conflicts at both the personal and the political levels, though of course conflicts between fundamental rights remain. Turning all conflicts into conflicts between rights removes this advantage.

Finally, there are difficulties with the ascription of rights in addition to those mentioned above in connection with the positive/negative distinction. Many of the issues involving genetics and the environment concern future generations, and Benatar suggests extending to future generations, for example, 'rights to a clean and safe environment'. This raises logical issues concerning rights which perhaps are best avoided by avoidance of rights-terminology itself. First, it may be asked, can non-existent beings have rights? Secondly, how are we to individuate these beings? The difficulty I have in mind here is what Derek Parfit has called the 'Non-identity problem'. The decisions we make concerning genetic engineering and so on are momentous, and will affect the identity of those who are born. Thus it is at least arguable that, even if the environment is seriously degraded as a result of those decisions, those who are born cannot claim that their rights have been violated, since the decisions themselves have benefited them. For without those decisions, they would not have existed.

Rights-violations are most obvious when they are most direct, and this gives rise to a further problem. If I torture you, then—on the assumption that it is indeed a right not to be tortured—I am undeniably violating your rights. But if I patent some set of human genes, perhaps as a spin-off from the Human Genome Project, and this plays a highly indirect part in the emergence of a genetic 'underclass' (p. 171), the matter is far less clear. The same would be true if I patented some genetically engineered crop that led to greater poverty in the developing world. In general, the harms and benefits that may arise from the biological revolution are hard to discern, and it is harder still to predict where and to whom they will fall. Vigilance and caution must, of course, be watchwords, but it does seem to me that the subtlety of what is at stake requires us to go beyond rights in our moral and political rhetoric.

Here I can note another point of agreement with Professor Benatar, who advocates that attention be paid not only to rights, but to 'human duties' (p. 186). Even here, however, it is important that morality alone should not be made to do all the work. The issues are too important and too urgent for that. Morality is useful, but it is not as efficient a motivator as self-interest. We must ensure that each individual is clear about the implications for his or her own well-being of the various policy options available, and seek to describe, as an ideal attractive to each individual, ways of life which are individually and collectively sustainable, and enable those who live them to play their part in securing a decent life for all.

Endnotes

Notes to Chapter 1

1. I understand that people have already had a child precisely for such a purpose.

2. Cf. William Gass's provocative essay, 'The Case of the Obliging Stranger' collected in his *Habitation of the Word* (New York: Simon and Schuster, 1985) for a brilliant defence of the significance of such spontaneous reactions, accompanied (unfortunately) by a radical pessimism as to the powers of ethical reflection.

3. It is important to make clear that all informed writers on the cloning issue point out that identical twins do not (and clones would not) have, for example, identical brains (most of the brain cells that are produced are eliminated in the growth process), depending on environmental contingencies (including internal environmental contingencies). Moreover, what cell assemblies get formed in the brain— the detailed 'wiring'—is heavily influenced by evolution. Thus identical twins do not and human clones would not have the same abilities and/or attitudes. And, while clones will look like the nucleus donor the resemblance will not be as strong as it is in the case of identical twins. (See Richard Gardner's remarks on this point in Chapter 2.)

4. *The Many Faces of Realism* (LaSalle, Ill.: Open Court, 1987).

5. In 'Moral Images and the Moral Imagination' published (in Spanish) as 'Imagenes Morales y Imaginacion Moral', *Dianoia*, 38 (1992), pp. 187–202.

6. This is a point that Ruth Anna Putnam stresses in the paper cited in the previous note. I am enormously indebted to

her reflections for helping me see more clearly just what I was trying to do with the notion of a moral image in *The Many Faces of Realism*.

7. *The Many Faces of Realism*, p. 79.
8. Richard Lewontin, 'Confusion about Cloning', *New York Review of Books*, 23 Oct. 1997, pp. 18–23.
9. Ibid. 21.
10. Here Lewontin is (unfortunately) mocking the language of *Cloning Human Beings: Report and Recommendations of the National Bioethics Advisory Commission* (Rockland, Md., June 1997).
11. *A Philosophy of Morals* (Oxford: Oxford University Press, 1990).
12. Daniel L. Kevles, reviewing Gunnar Broberg and Nils Roll-Hanson (eds.), *Eugenics and the Welfare State* in the *Times Literary Supplement*, 2 Jan. 1998, pp. 3–4.
13. Ibid. 3.
14. I use the Levinasian expression as a tribute to the philosopher who, more than any other, has taught us the moral importance of valuing *alterity*.

Note to the Response

1. A. Colman, 'Why Human Cloning would be Unhuman', *The Times*, 15 June 1998, p. 17.

Notes to Chapter 2

1. I. Wilmut *et al.*, 'Viable Offspring Derived from Fetal and Adult Mammalian Cells', *Nature*, 27 Feb. 1997.
2. T. Wakayama *et al.*, 'Full Term Development of Mice from Enucleated Oocytes Injected with Cumulus Cell Nuclei', *Nature*, 23 July 1998.

Notes to the Response

1. J. Commandon and P. De Fonbrune, 'Greffe nucleaire total, simple ou multiple, chez une amibe', *Compt. Rend. Soci. Biol.*, 130 (1939), pp. 744–8.

2. R. Briggs and T. J. King, 'Transplantation of Living Nuclei from Blastula Cells into Enucleated Frog Eggs', *Proceedings of the US National Academy of Science*, 38 (1952), pp. 455–63.

3. I. Wilmut, A. E. Schneike, J. McWhir, and K. H. S. Campbell, 'Viable Offspring Derived from Fetal and Adult Mammalian Cells', *Nature*, 385 (1997), pp. 810–13.

4. R. J. Sternberg and E. Grigorenko (eds.), *Intelligence, Hereditary and Environment* (Cambridge: Cambridge University Press, 1997); L. Wright, *Twins: Genes, Environment and the Mystery of Human Identity* (London: Weidenfeld & Nicholson, 1995).

5. D. K. Sokol, C. A. Moore, R. J. Rose, C. J. Williams, T. Reed, and J. C. Christian, 'Intrapair Differences in Personality and Cognitive Ability among Young Monozygotic Twins Distinguished by Chorion Type', *Behavior Genetics*, 25 (1995), pp. 457–66.

6. R. Ainslie, *The Psychology of Twinship* (Lincoln, Neb.: University of Nebraska Press, 1985).

7. P. Chichester, *A Hyphenated Life: The Bunker Twins Flourished in the N.C. Mountains* (Leisure Publishing Co., 1995).

8. D. J. P. Barker, 'Fetal Nutrition and Cardiovascular Disease in Later Life', *British Medical Bulletin*, 53 (1997), pp. 96–108.

Notes to Chapter 3

1. P. Billings in M. A. Rothstein and B. M. Knoppers, 'Legal Aspects of Genetics, Work and Insurance in North America and Europe', *European Journal of Health Law*, 3 (1996), pp. 143–61 at p. 161.

2. Trudo Lemmens, 'Genetics in Life, Disability and Additional Health Insurance in Canada: A Comparative

Legal and Ethical Analysis', in B. M. Knoppers (ed.), *Socio-Ethical Issues in Human Genetics* (Montreal: Yvon Blais, 1998).

3. H. Guay, B. M. Knoppers, and I. Panisset, 'La génétique dans les domaines de l'assurance et de l'emploi', *Revue du Barreau*, 52 (1992), pp. 185–343.

4. Rothstein and Knoppers, 'Legal Aspects'.

5. Human Genome Organization (HUGO), Hugo ethics committee, 'Statement on Sampling: Control and Access', *Genome Digest* (forthcoming).

6. Y. Sandberg, 'Genetic Information and Life Insurance: A Proposal for Ethical European Policy', *Social Science Medicine*, 40, 11 (1995), pp. 1553–4.

7. Lemmens, 'Genetics in Life Insurance'.

8. HUGO, ethics committee, 'Statement'.

Notes to the Response

1. Like Knoppers, I believe that it would be a mistake to devise genetic-specific anti-discrimination legislation; genetic conditions are best viewed as a discrete category within the general class of medical disability. Although I focus in this article on the case of bad genetic luck, my argument is intended to apply to bad luck in general with respect to one's health.

2. To this I add Eric Rakowski's proviso: 'To the extent that the unlucky did not willingly assume the risk of whatever disadvantages them . . .' See his *Equal Justice* (Oxford: Clarendon Press, 1991), p. 2. I do not think, as does Hillel Steiner (see Chapter 6), that parents could ever be fairly held responsible for their offspring's bad genetic luck nor that they should be. I hold the former view because most of us will be like the 'polys' in the hypothetical society discussed below, namely, people who have genetic predispositions and susceptibilities to certain conditions that will

only become manifest if we are exposed to disease-aggravating environments. Environmental influences on gene expression are so numerous and varied that it is doubtful their effects on many genes will ever be known with much precision, if they will be known at all. Thus it will never be possible to test gametes and/or zygotes for genetic defects in the way that Steiner suggests. His depiction of what he calls the 'post-revolutionary' world is inaccurate. The reasons I disagree that parents should be held liable if they bring children into existence who possess defective genes are discussed in my *Genetic Justice* (Oxford: Oxford University Press, forthcoming).

3. I do not, in this comment, provide an account of what form such a scheme should take. The one I have in mind is similar to that discussed in detail by Ronald Dworkin in his 'What is Equality? Part II: Equality of Resources', *Philosophy and Public Affairs*, 10 4 (1981), pp. 283–345.

4. Rates could legitimately differ for reasons connected to income and resource holdings.

5. In the hypothetical society under discussion we must assume that there are enough low risk 'polys' to make this a financially viable strategy for insurers. In the absence of enough low-risk insurers participating in the private insurance market premiums charged to those at high risk (even for 'no questions asked policies') might be unaffordable.

6. It is also eminently arguable that state health insurance would be more cost effective.

Notes to Chapter 4

1. Wilmut *et al.*, 'Viable Offspring Derived from Fetal and Adult Mammalian Cells', *Nature*, 27 Feb. 1997.

2. See *Cloning Human Beings: Report and Recommendations of*

the National Bioethics Advisory Commission (Rockville, Md., June 1997).

3. From President Clinton's weekly radio broadcast reported in *Bioworld Today*, 9/7 (13 Jan. 1998). Interestingly, the National Bioethics Advisory Commission (n. 2 above) stated that it was unethical *because* unsafe. Either Clinton misread his advisors' report or decided to add 'morally unacceptable' on top of the fact that it was untested and unsafe rather than simply because it was untested and unsafe.

4. Reported in *BioCentury: The Bernstein Report on BioBusiness*, 19 Jan. 1998.

5. WHO Press Release (WHO/20, 11 Mar. 1997).

6. WHO document (WHA50.37, 14 May 1997). Despite the findings of a Meeting of the Scientific and Ethical Review Group (see n. 1) which recommended that 'the next step should be a thorough exploration and fuller discussion of the [issues]'.

7. UNESCO Press Release No. 97–29.

8. UNESCO *Universal Declaration on the Human Genome and Human Rights*, published by UNESCO, 3 Dec. 1997.

9. Michel Revel, 'Questioning the Ban on Human Cloning', *The Scientist*, 19 Jan. 1998.

10. Ibid.

11. Federico Mayor, 'Devaluing the Human Factor', *The Times Higher*, 6 Feb. 1998, p. 13.

12. The European Parliament, Resolution on Cloning, Motion dated 11 Mar. 1997, passed 13 Mar. 1997.

13. Axel Kahn, 'Clone Mammals . . . Clone Man', *Nature*, 386, 13 Mar. 1997, p. 119.

14. See my 'Is Cloning an Attack on Human Dignity?', *Nature*, 387, 19 June 1997, p. 754.

15. *Opinion of the Group of Advisers on the Ethical Implications of*

Biotechnology to the European Commission, No. 9, 28 May 1997. Rapporteur Dr Anne McClaren.

16. Axel Kahn, *Nature*, 388, 24 July 1997, p. 320.

17. Richard Lewontin, 'Confusion about Cloning', *New York Review of Books*, 23 Oct. 1997, pp. 18–23.

18. In my 'Rights and Human Reproduction', in John Harris and Søren Holm (eds.), *The Future of Human Reproduction: Choice and Regulation* (Oxford: Oxford University Press, 1998).

19. Ibid.

20. I realize Putnam says nothing about punishment, but by saying that Nazism might easily follow were cloning permitted, Putnam is certainly giving support to those who would outlaw human cloning.

21. Ibid.

22. My added emphasis.

23. Leaving aside, for the sake of the moral image, the interests of the child.

24. And, for good measure, perhaps we should use legislation to prevent any family having an established religion.

25. Another argument for the surprise factor in cloning!

26. Mayor, 'Devaluing'.

27. *The Times Higher*, 30 Jan. 1998, pp. 18–19.

28. It is unlikely that 'artificial' cloning would ever approach such a rate on a global scale and we could, of course, use regulative mechanisms to prevent this without banning the process entirely. I take this figure of the rate of natural twinning from Keith L. Moore and T. V. N. Persaud, *The Developing Human*, 5th edn. (Philadelphia; W. B. Saunders, 1993). The rate mentioned is 1 per 270 pregnancies.

29. Mitochondrial DNA individualizes the genotype even of clones to some extent. The mitochondria are particles of DNA present in each egg cell and are derived from the mother of that egg. They are additional to the 46

chromosomes that make up the genome which is cloned using nuclear substitution.

30. Although of course there would be implications for criminal justice since clones could not be differentiated by so called 'genetic fingerprinting' techniques.

31. David Hume in his *A Treatise of Human Nature* (1738). Contemporary philosophers who have flirted with a similar approach include Stuart Hampshire, see e.g. his *Morality and Pessimism—The Leslie Stephen Lecture* (Cambridge: Cambridge University Press, 1972), and Bernard Williams, see his 'Against Utilitarianism' in B. Williams and J. J. C. Smart (eds.), *Utilitarianism For and Against* (Cambridge: Cambridge University Press, 1973). I first discussed the pitfalls of olfactory moral philosophy in my *Violence and Responsibility* (London: Routledge & Kegan Paul, 1980).

32. Mary Warnock, 'Do Human Cells Have Rights?', *Bioethics*, 1/1 (Jan. 1987), p. 8.

33. Leon R. Kass, 'The Wisdom of Repugnance', *The New Republic*, 2 June 1997, pp. 17–26. The obvious erudition of his writing leads to expectations that he might have found feelings prompted by more promising parts of his anatomy with which to entertain us.

34. In a letter to Humphrey House, 11 Apr. 1940: *The Collected Essays, Journalism and Letters of George Orwell*, i (Harmondsworth; Penguin, 1970), p. 583. See my more detailed discussion of the problems with this type of reasoning in *Wonderwoman and Superman: The Ethics of Human Biotechnology* (Oxford: Oxford University Press, 1992), ch. 2.

35. See n. 3 above.

36. We await the report of the United Kingdom Human Genetics Advisory Commission.

37. See n. 9 above.

38. Histocompatible simply means 'compatible tissue'; the key

point is that organs must be sufficiently similar to avoid dangers of rejection when implanted in a host.

39. See J. M. W. Slack *et al.*, 'The Role of Fibroblast Growth Factors in Early Xenopus Development', in *Biochemical Society Symposium*, 62 pp. 1–12.

40. When such procedures can be considered sufficiently safe.

41. Ronald Dworkin, *Life's Dominion* (London: HarperCollins, 1993), p. 148. See also John A. Robertson, *Children of Choice* (Princeton, NJ: Princeton University Press, 1994), esp. ch. 2.

42. Ronald Dworkin, *Freedom's Law* (Oxford: Oxford University Press, 1996), pp. 104–5.

43. Ibid. 237–8.

44. Ronald Dworkin has produced an elegant account of the way the price we should be willing to pay for freedom might or might not be traded off against the costs. See his *Taking Rights Seriously* (London: Duckworth, 1977), ch. 10, and his *A Matter of Principle* (Cambridge, Mass.: Harvard University Press, 1985), ch. 17.

45. Dworkin, *Life's Dominion*, pp. 167–8.

46. *State of Washington et al. Petitioners* v. *Glucksberg et al.*, and *Vacco et al.* v. *Quill et al.* argued 8 Jan. 1997.

47. Mayor, 'Devaluing'.

48. Some of the material in this lecture was presented to the *UNDP/WHO/World Bank Special Programme of Research, Development and Research Training in Human Reproduction Review Group Meeting*, Geneva, 25 Apr. 1997, and to a hearing on cloning held by the European Parliament in Brussels, 7 May 1997. I am grateful to participants at these events for many stimulating insights. The general issues raised by cloning were discussed in a special issue of the *Kennedy Institute of Ethics Journal*, 4/3 (1994), and in my *Wonderwoman and Superman: The Ethics of Human Biotechnology* (Oxford: Oxford University Press, 1992), esp. ch. 1.

Versions of the ideas expressed here have appeared in my 'Goodbye Dolly: The Ethics of Human Cloning', *Journal of Medical Ethics*, 23/6 (1997), pp. 353–61, and in my 'Cloning and Human Dignity', in *Cambridge Quarterly of Healthcare Ethics*, 7 (1998), pp. 163–7.

Notes to the Response

1. One important practical consideration that I do not discuss here is the issue of safety. See Alan Colman's comment on Putnam in Chapter 1 for discussion of this point.
2. The Human Fertilisation and Embryology Act 1990.
3. *Report of the Committee of Inquiry into Human Fertilisation and Embryology* (1984), Cmnd. 9314.
4. Para. 11.17.
5. Para. 11.30.
6. Ss. 3, 15.
7. S. 13(5).
8. The need to distinguish between therapeutic and reproductive uses of 'cloning' technology was recently affirmed in the *Human Fertilisation and Embryology Authority/Human Genetics Advisory Commission Joint Consultation Paper on Cloning* (1998). Neither may be legal as the law stands.
9. It should also be pointed out that current rules governing scientific and medical progress are of advantage to scientists and medics. Public reassurance that research on embryos is being monitored and is not out of control creates an atmosphere of confidence and respectability in which research may take place and gain public funding. Furthermore, detailed regulation, such as the guidelines laid down in the HFEA Code of Practice for clinicians, provides some protection against malpractice litigation by disappointed patients.

Notes to Chapter 5

1. Benno Muller-Hill, *Murderous Science: Elimination by Scientific Selection of Jews, Gypsies, and Others, Germany 1933–1945*, trans. George R. Fraser (Oxford: Oxford University Press, 1988), p. 10.
2. Cf. Robert N. Proctor, *Racial Hygiene Medicine Under the Nazis* (Cambridge, Mass.: Harvard University Press, 1988).
3. *Hitler's Table Talk*, intro. Hugh Trevor-Roper (Oxford: Oxford University Press, 1988), pp. 396–7.
4. Quoted in Robert Jay Lifton, *The Nazi Doctors: A Study in the Psychology of Evil* (London: Macmillan, 1986), p. 16.
5. Quoted in Muller-Hill, *Murderous Science*, pp. 10, 12.
6. Quoted in Rual Hilberg, *The Destruction of the European Jews* (New York, Holmes & Meier, 1985), p. 287.
7. *Hitler's Table Talk*, p. 332.
8. Quoted in Jeremy Noakes and Geoffrey Pridham, *Nazism 1919–1945: A Documentary Reader* (Exeter: University of Exeter, 1983–8), iii. 1014–15.
9. Quoted in Muller-Hill, *Murderous Science*, p. 14.
10. Quoted in Lifton, *Nazi Doctors*, p. 16.
11. Richard Dawkins, *The Selfish Gene* (Oxford: Oxford University Press, 1976), p. 215.

Notes to Chapter 6

1. See, for example, the report in the *Independent*, 4 Mar. 1997, entitled 'Labor redefines the age of innocence': 'Children between 10 and 13 should be deemed "capable of evil", Jack Straw, the shadow Home Secretary declared yesterday . . . as he urged the scrapping of the presumption against [their] criminal responsibility.'
2. I do, however, address this issue in *An Essay on Rights*, (Oxford: Blackwell, 1994), pp. 245–6, and in 'Working Rights', sect. 3, in Matthew Kramer, Nigel Simmonds, and Hillel Steiner, *A Debate Over Rights: Philosophical Enquiries*

(Oxford: Oxford University Press, 1998), where I argue that minors do *not* have rights.

3. A good indication of the aforesaid pressure is that, in his rather spare jurisprudential classification of basic types of right—as either property or contract—Kant found it necessary to devise a third category, 'Rights to Persons Akin to Rights to Things', to cover parental rights with regard to minors; cf. Immanuel Kant, *The Metaphysics of Morals*, trans. Mary Gregor (Cambridge: Cambridge University Press, 1991), pp. 98–100.

4. Cf. Kant, *Metaphysics of Morals*, p. 63. A thorough explication of the concept of self-ownership is to be found in G. A. Cohen, *Self-Ownership, Freedom and Equality* (Cambridge: Cambridge University Press, 1995), ch. 9; see also my *An Essay on Rights*, pp. 231–48.

5. On the significance of, and conditions for, the absence of duty-conflicts—the mutual consistency of rights—see my *An Essay on Rights*, pp. 55–108, 224–8, and 'Working Rights', sect. 4.

6. Cf. the *Independent*, 9 Dec. 1997: 'Blair puts parents in firing line over school attendance': 'Parents could be fined £1000 if they fail to make sure that their children are attending school, the Government warned yesterday'. Also the *Independent*, 2 Dec. 1996: 'Pupils sue schools for bad education'; a lawyer for the pupils argued that 'If you have a car where the brakes fail there are victims; similarly, if you have a school which fails its pupils, there are victims'. The case followed one in which an out-of-court settlement resulted in a '20-year-old [accepting] £30,000 damages after claiming compensation for bullying at school . . . Both cases have been made possible because of government policy which requires inspectors to say when schools are failing.'

7. Cf. David Heyd, *Genethics* (Berkeley: University of California Press, 1992), pp. 26–38: 'Tort is universally defined

as some sort of *worsening* in the condition of a person
(damages being an attempt to compensate for exactly that
difference between the person's condition before the harm
was done and that following it). But in [the case of genetic
defects], the only condition that can be the subject of such
a comparison and a point of reference for the assessment
of appropriate compensation is nonexistence . . . The incli-
nation to avoid comparisons in wrongful life cases is not
only motivated by the fact that nonexistence is not a state
that can be given a value, but also because it is not a
state that can be *attributed* to a subject. It is hardly a "state"
at all' (pp. 29–30).

8. John Locke, *Two Treatises of Government*, ed. Peter Laslett
(Cambridge: Cambridge University Press, 1967), second
treatise, ch. V.

9. In *An Essay on Rights*, chs. 6(D), 7(A), I argue that this
duty, along with the duty to respect others' self-ownership,
are both directly implied by each person's having a foundational right to equal freedom.

10. Though perhaps an unconditional basic initial endowment,
being less paternalistic, would be a more Kantian or liberal
instantiation of this entitlement. An excellent account of
the justice-based case for unconditional basic income is
offered in Philippe Van Parijs, *Real Freedom for All* (Oxford:
Oxford University Press, 1995).

11. Indeed, since the likely impact of an egalitarian redistribution of all natural resource values—on a global scale—
would be considerably to reduce interpersonal inequalities
of wealth generally, we might expect children's ability-level
differentials to be still further reduced. For it is reasonable
to suppose that the effect of that former reduction on
opportunity costs, including those of producing a given
level of ability in their children, would be downward for
many more (largely poorer) persons than those (largely

wealthier) for whom it would be upward. Having command over more silver production factors to combine with their children's genetic endowments, more people would have to sacrifice less to raise children who were healthier, more skilled, and better informed. In *An Essay on Rights*, pp. 249–58, 273, I argue that the full value of dead persons' estates is also justly subject to such egalitarian redistribution. Still further reduction in global wealth inequalities might be expected from measures to secure restitution for past injustices, including the injustice of having withheld the foregoing redistributive entitlements.

12. Since, on this reading, my genetic characteristics are neither conjunctively nor disjunctively constitutive of my identity, perhaps what we have here is a fairly natural (*sic*) approximation to the otherwise elusive notion of the Kantian 'noumenal self'.

13. An incidental but important feature of this finding is its illumination of a general shortcoming in many of those currently prominent justice theories which are associated with the Kantian ends–means injunction and which take equality of opportunity as their central distributive principle. John Roemer has usefully explicated this conception of equality as follows: 'At a philosophical level, many people associate egalitarianism, and the policies of the welfare state in particular, with a view that society will indemnify citizens against all major harms, relieving them of the personal responsibility to make their lives go right. I shall not defend this kind of unqualified egalitarianism, which does not hold individuals responsible for their choices. Equality of opportunity, in contrast, is a view that society (the government) must level the playing field, but that after that, individuals should suffer or enjoy the consequences of their own choices. The question becomes: Exactly what is required to level the playing field? . . . Let us say that a

person's actions or behavior are determined by two kinds of cause: *circumstances* beyond her control, and *autonomous choices* within her control . . . A particular action a person takes, and its associated consequences, are caused by a highly complex combination of circumstances and autonomous choices. I say that equality of opportunity has been achieved among a group of people if society indemnifies persons in the group against that part of consequences they suffer due to circumstances and brute luck, but does not indemnify them against that part of consequences due to autonomous choice. Thus the purpose of an equal-opportunity policy is to equalize outcomes in so far as they are the consequences of causes beyond a person's control, but to allow differential outcomes in so far as they result from autonomous choice: 'Equality of Opportunity: Theory and Examples' (mimeo, University of California at Davis, June 1995), pp. 2–5; also *Equality of Opportunity* (Cambridge, Mass.: Harvard University Press, 1998). Many such equal-opportunity theories, however, suffer from their inconsistent application of that personal responsibility requirement, in so far as they assimilate—as due to 'circumstance'—those adverse consequences which persons incur at the hands of others, to those which are imputable to brute luck or what I here refer to as nature. In so doing, they are led to infer that compensatory liability for the former, as well as the latter, lies with society rather than the particular individuals responsible for those adversities: that is, they embrace social indemnification of harmers; cf. my 'Choice and Circumstance', *Ratio*, 10 (1997), pp. 296–312.

14. And there are good reasons to believe that Kantian—or, more generally, liberal—conceptions of justice cannot allow the enforceability of moral values other than justice.

Notes to Chapter 7

1. E. Hobsbawm, *The Age of Extremes* (New York: Pantheon, 1994).

2. S. R. Benatar, 'Africa and the World', *South African Medical Journal*, 84 (1994), pp. 723–6; D. Logie and S. R. Benatar, 'Africa in the 21st Century', *British Medical Journal*, 315 (1997), pp. 1444–6; S. R. Benatar, 'Black Africa: Between Hope and Despair', in R. Buckley (ed.), *Understanding Global Issues*, 97/6 (1997); S. R. Benatar, 'Towards Social Justice in the New South Africa', *Medicine, Conflict, and Survival*, 13 (1997), pp. 229–39.

3. S. R. Benatar, 'Prospects for Global Health: Lessons from Tuberculosis', *Thorax*, 50 (1995), pp. 487–9.

4. Benatar, 'Prospects for Health'; J. Grange and A. Zumla, 'Tuberculosis: An Epidemic of Injustice', *Journal of Royal College of Physicians London*, 31 (1997), pp. 637–40.

5. United Nations Development Programme, *Human Development Report 1991* (Oxford: Oxford University Press, 1991); A. Gilbert, *An Unequal World: The Link between Rich and Poor Nations*, 2nd edn. (Walton-on-Thames: Nelson, 1992); R. Broad and J. Cavanagh, 'Don't Neglect the Impoverished South', *Foreign Policy*, 101 (1995), pp. 18–35.

6. *Human Development Report*.

7. H. C. J. Van Rensburg and S. R. Benatar, 'The Legacy of Apartheid in Health and Health Care', *South African Journal of Sociology*, 24 (1993), pp. 99–111.

8. J. Stone, *Racial Conflict in Contemporary Society* (Cambridge, Mass.: Harvard University Press, 1985); G. Teeple, *Globalisation and the Decline of Social Reform* (Atlantic Highlands, NJ: Humanities Press, 1995); I. Wallerstein, 'America and the World: Yesterday, Today and Tomorrow', *Theory and Society*, 21 (1992), pp. 1–28; R. Heilbroner, *21st-Century Capitalism* (New York: W. W. Norton, 1993).

9. *Human Development Report*; Broad and Cavanagh, 'Don't Neglect'; Teeple, *Globalisation*; Wallerstein, 'America'.

10. Teeple, *Globalisation*; Wallerstein, 'America'; Heilbroner, *21st-Century Capitalism*.

11. S. R. Benatar, 'Global Disparities in Health and Human Rights: A Critical Commentary', *American Journal of Public Health*, 88 (1998), pp. 295–300; S. P. Riley (ed.), *The Politics of Global Debt* (London: Macmillan, 1993); S. R. Benatar, 'World Health Report 1996: Some Millennial Challenges', *Journal of Royal College of Physicians London*, 31 (1997), pp. 456–7.

12. J. K. Galbraith, *The Good Society: The Human Agenda* (London: Sinclair Stevenson, 1996).

13. Broad and Cavanagh, 'Don't Neglect'.

14. Hobsbawm, *Age of Extremes*; Benatar, 'Black Africa'; K. Booth, 'Human Wrongs and International Relations', *International Affairs*, 71 (1995), pp. 103–26; A. J. McMichael, *Planetary Overload: Global Environmental Change and the Health of the Human Species* (London: Cambridge University Press, 1993); *Our Global Neighbourhood: Report of the Commission on Global Governance*, Co-Chairmen I. Carlsson and S. Ramphal (Oxford: Oxford University Press, 1995); *Health and Environment in Sustainable Development: Five Years after the Earth Summit* (Geneva: World Health Organisation, 1997).

15. Teeple, *Globalisation*.

16. R. G. Wilkinson, *Unhealthy Societies: The Afflictions of Inequality* (New York: Routledge, 1996).

17. G. Khushf, 'Ethics, Policy and Health Care Reform', *Journal of Medicine and Philosophy*, 19 (1994), pp. 397–405.

18. A. Chapman (ed.), *Health Care Reform: A Human Rights Approach* (Washington DC: Georgetown University Press, 1994); S. R. Benatar, 'Just Healthcare beyond Individualism:

Challenges for North American Bioethics', *Cambridge Quarterly of Healthcare Ethics*, 6 (1997), pp. 397–415.

19. A. Chetley, *A Healthy Business? World Health and the Pharmaceutical Industry* (London: Zed Books, 1990).

20. R. G. Spece, D. S. Shim, and A. E. Buchanan, *Conflicts of Interest in Clinical Practice and Research* (New York: Oxford University Press, 1996).

21. Benatar, 'Just Healthcare'.

22. V. Sidel, 'The International Arms Trade and its Impact on Health', *British Medical Journal*, 311 (1995), pp. 1677–80.

23. M. J. Toole and R. J. Waldman, 'Refugees and Displaced Persons: War, Hunger and Public Health', *Journal of the American Medical Association*, 270 (1993), pp. 600–5; A. P. Schmidt (ed.), *PIOOM Newsletter and Report*, 7/1 (1995).

24. Logie and Benatar, 'Africa'; S. Gervasi and S. Wong, 'The Reagan Doctrine and Destabilisation of Southern Africa', in A. George (ed.), *Western State Terrorism* (Cambridge: Quality Press), pp. 212–52.

25. Logie and Benatar, 'Africa'; Benatar, 'Black Africa'.

26. 'Torture and the Medical Profession: Proceedings of an International Symposium', *Journal of Medical Ethics*, 17 supp. (1991), pp. 1–64; M. Basoglu, 'Prevention of Torture and Care of Survivors', *Journal of the American Medical Association*, 270 (1993), pp. 605–11; V. Iacopino, M. Heisler, S. Pishevar, and R. H. Kirschner, 'Physician Complicity in Misrepresentation and Omission of Evidence of Torture in Post-Detention Medical Examinations in Turkey', *Journal of the American Medical Association*, 276 (1996), pp. 396–402.

27. A. Cassese, *Inhuman States: Imprisonment, Detention and Torture in Europe Today* (Cambridge: Polity Press, 1996).

28. Teeple, *Globalisation*; Wallerstein, 'America'; Heilbroner, *21st-Century Capitalism*; Galbraith, *Good Society*; Booth, 'Human Wrongs'; Benatar, 'Global Disparities'; Riley, *Politics of Debt*; S. George and F. Fabelli, *Faith and Credit: The*

World Bank's Secular Empire (London: Penguin, 1994); J. Donnelly, *Universal Human Rights in Theory and Practice* (London: Cornell University Press, 1989); S. Hoffman, 'The Crisis of Liberal Internationalism', *Foreign Policy*, 98 (1995), pp. 159–77; Benatar, 'World Health'.

29. A. A. Mazrui, *The Africans: A Triple Heritage* (London: BBC, 1986); C. Achebe, *Hopes and Impediments: Selected Essays 1965–87* (Oxford: Heinemann, 1988); O. Ngugi Wa Thiong', *Moving the Centre: The Struggle for Cultural Freedoms* (London: James Currey, 1993).

30. Benatar, 'Africa'; Logie and Benatar, 'Africa'; Benatar, 'Black Africa'; George and Fabelli, *Faith and Credit*; A. Pettifor, *Debt, the Most Potent Form of Slavery* (London: Christian Aid, 1996).

31. Pettifor, *Debt*.

32. Benatar, 'Africa'; Logie and Benatar, 'Africa'; Benatar, 'Black Africa'.

33. M. Michaels, 'Retreat from Africa', *Foreign Affairs*, 72 (1992/3), pp. 93–108.

34. M. Hales, *Science or Society?* (London: Pan, 1982).

35. A. K. Gupta, 'Ethical Dilemmas in Conservation of Biodiversity: Towards Developing Globally Acceptable Ethical Guidelines', *Eubios Journal of Asian and International Bioethics*, 5/2 (1995), pp. 40–6; H. Harrada, 'Advancement of Plant Breeding Techniques: Scientific, Social and Global Impact', *Eubios Journal of Asian and International Bioethics*, 6/5 (1997), pp. 131–4.

36. T. Gillow, 'The Psychological, Social and Economic Consequences of Blinding Soldiers', *Eubios Journal of Asian and International Bioethics*, 13 (1997), pp. 327–32.

37. W. M. Arkin, 'Acoustic Anti-Personnel Weapons: An Inhumane Future?', *Eubios Journal of Asian and International Bioethics*, 13 (1997), pp. 314–26.

38. W. Barnaby, 'Biological Weapons: An Increasing Threat', *Medicine, Conflict and Survival*, 13 (1997), pp. 301–13.

39. H. Nickens, 'The Genome Project and Health Services for Minority Populations', in T. H. Murray, M. A. Rothstein, and R. F. Murray (eds.), *The Human Genome Project and the Future of Health Care* (Bloomington, Ind.: Indiana University Press, 1996), pp. 59–78.

40. Donnelly, *Universal Human Rights*.

41. McMichael, *Planetary Overload*; *Our Global Neighbourhood*; *Health and Environment*.

42. J. K. Galbraith, *The Culture of Contentment* (Boston: Houghton Mifflin, 1992).

43. Galbraith, *Good Society*; M. J. Lerner, *The Belief in a Just World: A Fundamental Delusion* (New York: Plenum Press, 1980); A. Kleinman, V. Das, and M. Lock (eds.), 'Social Suffering: Introduction', *Daedalus: Proceedings of the American Academy of Arts and Sciences*, 125/1 (1996), pp. 11–20; S. R. Benatar, 'War or Peace and Development', *Medicine, Conflict and Survival*, 13 (1997), pp. 125–34.

44. S. Mulhall and A. Swift, *Liberals and Communitarians*, 2nd edn. (London: Blackwell, 1996); N. L. Rosenblum, *Liberalism and the Moral Life* (Cambridge, Mass.: Harvard University Press).

45. Mulhall and Swift, *Liberals*; Rosenblum, *Liberalism*; S. Benhabib, *Situating the Self: Gender, Community and Postmodernism in Contemporary Ethics* (New York: Routledge, 1992).

46. Hoffman, 'Crisis'.

47. Gilbert, *Unequal World*; Galbraith, *Good Society*; Riley, *Politics of Debt*; George and Fabelli, *Faith and Credit*; Donnelly, *Universal Human Rights*; Hoffman, 'Crisis'.

48. Teeple, *Globalisation*.

49. N. Gardels, 'The Post Atlantic Capitalist Order', *New Perspectives Quarterly*, 2–3 (1993), pp. 2–3.

50. R. Heilbroner, 'The Rest of the World off the Track:

Growth and the Lumpen Planet', *New Perspectives Quarterly*, 2–3 (1993), pp. 48–53.

51. Heilbroner, *21st-Century Capitalism*; Galbraith, *Good Society*.

52. A. R. Chapman, 'Reintegrating Rights and Responsibilities', in K. W. Hunert and T. C. Mack (eds.), *International Rights and Responsibilities for the Future* (Westport, Conn.: Praeger, 1996).

53. R. E. Howard, 'Monitoring Human Rights: Problems of Consistency', *Ethics and International Affairs*, 4 (1990), pp. 33–71.

54. Howard, 'Monitoring'; Amnesty International, *United States of America: Human Rights Violations: A Summary of Amnesty International's Concerns* (New Haven: UK Amnesty International, 1995).

55. McMichael, *Planetary Overload*; *Our Global Neighbourhood*.

56. G. Hardin, 'The Tragedy of the Commons', *Science*, 162 (1968), pp. 1243–8.

57. McMichael, *Planetary Overload*; *Our Global Neighbourhood*; *Health and Environment*.

58. Benatar, 'Black Africa'; G. K. Heillener, 'The IMF, the World Bank and Africa's Adjustment and Debt Problems', *World Development*, 20 (1992), pp. 779–92.

59. Wallerstein, 'America'; Booth, 'Human Wrongs'.

60. Hoffman, 'Crisis'.

61. Teeple, *Globalisation*; Wallerstein, 'America'; Heilbroner, *21st-Century Capitalism*; Galbraith, *Good Society*; Galbraith, *Culture of Contentment*.

62. *Human Development Report 1991*.

63. Gilbert, *Unequal World*; Broad and Cavanagh, 'Don't Neglect'; Galbraith, *Good Society*.

64. Logie and Benatar, 'Africa'.

65. Hoffman, 'Crisis'; R. Attfield and B. Wilkins, *International Justice and the Third World* (London: Routledge, 1992).

66. Heilbroner, *21st-Century Capitalism*.

67. Heilbroner, *21st-Century Capitalism*; Galbraith, *Good Society*; Galbraith, *Culture of Contentment*.

68. Galbraith, *Good Society*.

69. (i) It must not be aggressive or have expansionist aims, (ii) it must have a system of laws guided by a 'common good' conception of justice that takes into account people's essential interests, which include the right to means of subsistence and security, to liberty, personal property, and to formal equality, as expressed by the rules of natural justice, and (iii) it must respect basic human rights: J. Rawls, 'The Law of Peoples', in S. Shute and S. Hurley (eds.), *On Human Rights: The Oxford Amnesty Lectures 1993* (New York: Basic Books, 1993), pp. 41–82.

70. Declaration of Human Duties: A Code of Ethics and Shared Responsibilities, 'Correspondence' to the Committee on Human Rights, US National Academy of Sciences, National Academy of Engineering, Institute of Medicine, autumn 1995, p. 6.

71. Booth, 'Human Wrongs'; V. Stambolovic, 'Human Rights and Health within the Dominant Paradigm', *Social Science Medicine*, 42 (1996), pp. 1–303; S. R. Benatar, 'Key Ethical Dimensions of the Renewal Process at the Global Level: Streams of Global Change', in Z. Bankowski, J. H. Bryant, and J. Gallagher (eds.), *Ethics, Equity and Health for All* (Geneva: CIOMS, 1977); S. R. Benatar, 'Millennial Challenges for Modernity and Medicine', *Journal of Royal College of Physicians*, 32 (1998), pp. 160–5.

72. Benatar, 'Global Disparities'.

73. Achebe, *Hopes and Impediments*.

Index